£ 3.30

Alex Strachan

DETAILED SOLUTIONS TO A-LEVEL QUESTIONS

GW00480777

D. Murray BSc, CertEd
Head of Mathematics
Dover Grammar School for Boys

D. Benjamin MSc
Assistant Head of Mathematics
Dover Grammar School for Boys

Stanley Thornes (Publishers) Ltd

3

First published in 1986 by
Stanley Thornes (Publishers) Ltd
Old Station Drive
Leckhampton Road
CHELTENHAM GL53 0DN

British Library Cataloguing in Publication Data

Murray, D.
    Detailed solutions to A-level questions in pure
    mathematics.
    1. Mathematics — Examinations, questions, etc.
    I. Title   II. Benjamin, D.
    510′.76   QA43

    ISBN 0–85950–522–7

Typeset by Tech-Set, Gateshead, Tyne & Wear in 9/10pt Press Roman.
Printed and bound in Great Britain at The Bath Press, Avon.

CONTENTS

PREFACE

This book, together with the one on Applied Mathematics, is primarily aimed at students of the London University (Syllabus B) A-level in Pure and Applied Mathematics.

The idea for the books arose from the need of our own students. Whilst working through their booklets of past papers they were often unable to seek assistance with problems, either because they were working at home or involved in private study at school. It was therefore necessary to produce detailed solutions which were easy to follow. Consequently at all stages we have clearly shown the method being used and for fluency of reading we have also ensured that all algebraic steps are shown in full. This approach has produced some lengthy solutions but we have found them to be instructive to our students. The methods are printed in italics, and whilst students will not be expected to write their own solutions in such detail, they do indicate that a plan of approach is important when attacking problems. Large diagrams, clearly showing all relevant forces and any given information, also aid the solutions.

The question papers can be obtained from the University of London Publications Office, 52 Gordon Square, London, WC1 E7HU.

We would like to thank the following for their invaluable help:
Sandra Bateman, Patsy Ford, Jean Ifield and Di Mayes for their typing of the manuscript; Julie Liddon for her secretarial work; Bryan Quinn for his criticism and advice; and Simon McBride, Philip Keates, Andrew Kenchington, John Monger and David Wouldham, from our Sixth Form, for their proof reading.

The University of London University Entrance and School Examinations Council accepts no responsibility whatsoever for the accuracy or method of working in the solutions given.

The authors take sole responsibility for all solutions.

DAVID MURRAY
DAVID BENJAMIN

NOTATION AND ABBREVIATIONS

Notation used:

$a = b$	a is equal to b
$a \neq b$	a is not equal to b
$a \equiv b$	a is equivalent to b
$a \approx b$	a is approximately equal to b
$\pm a$	plus or minus a
$a > b$	a is greater than b
$a \geqslant b$	a is greater than or equal to b
$a < b$	a is less than b
$a \leqslant b$	a is less than or equal to b
$a!$	factorial a (e.g. $5! \equiv 5 \times 4 \times 3 \times 2 \times 1$)
$\lvert a \rvert$	modulus of a (e.g. $\lvert 5 \rvert = 5$, $\lvert -5 \rvert = 5$)
$\mathbb{R}$	the set of real numbers
$a \in \mathbb{R}$	a is an element of the set of real numbers
$\{a : a \in A\}$	the set of elements of A such that $a \in A$
$A \cup B$	the union of sets A and B
$A \cap B$	the intersection of sets A and B
$\overline{A}$	all elements not in set A
$\log x$	the logarithm of x, base 10
$\ln x$	the natural logarithm of x, base e
$\displaystyle\sum_{r=1}^{n} u_r$	$u_1 + u_2 + \cdots + u_{n-1} + u_n$
^nC_r	the number of ways of choosing r objects from n different objects, i.e. $\dfrac{n!}{(n-r)!r!}$
$f : x \to y$	the function f which maps x on to y
$f(x)$	the image of x under the function f
$f^{-1}(x)$	the inverse of the function f
$fg(x)$	the image of $g(x)$ under the function f
$\arcsin x$ or $\sin^{-1} x$	the inverse of the function $\sin x$
$\arctan x$ or $\tan^{-1} x$	the inverse of the function $\tan x$
$f'(x)$	the derivative of $f(x)$ with respect to x
$\dfrac{d^n y}{dx^n}$	the nth derivative of y with respect to x
δx	a small amount of x
$x \to +\infty$	x tends to 'positive infinity'
$x \to -\infty$	x tends to 'negative infinity'
$x \to a$	x tends to a constant a
$\overrightarrow{OA}$	the vector from O to A
$\mathbf{r}$	the vector $\mathbf{r}$
$\lvert \mathbf{r} \rvert$	the length of the vector $\mathbf{r}$
$\hat{\mathbf{r}}$	the unit vector in a direction $\mathbf{r}$

Abbreviations used:

$A\widehat{O}B$	the angle AOB
L.H.S.	left-hand side
R.H.S.	right-hand side

January 1983

Let $f(x) \equiv x^3 + px + r$.

By the *Remainder Theorem* if $f(x)$ has a *remainder* of -9 when divided by $(x + 1)$ then

$$f(-1) = -9$$

Therefore

$$f(-1) = -1 - p + r = -9$$
$$\Rightarrow \quad -p + r = -8 \qquad\qquad [1]$$

Similarly if $f(x)$ has a *remainder* of -1 when divided by $(x - 1)$ then

$$f(-1) = -1$$

Therefore

$$f(1) = 1 + p + r = -1$$
$$\Rightarrow \quad p + r = -2 \qquad\qquad [2]$$

Adding [1] to [2] gives

$$2r = -10$$
$$\Rightarrow \quad r = -5$$

Substituting $r = -5$ into [2] gives

$$p - 5 = -2$$
$$\Rightarrow \quad p = 3$$

Therefore the constants p and r are -3 and -5 respectively.

Let $f(x) = |x + 4|$ and $g(x) = |x + 3|$.

We will sketch the two functions $f(x)$ and $g(x)$ on the same pair of axes and where $f(x)$ lies below $g(x)$ will be the solution to the inequality

$$|x + 4| < |x + 3|$$

$f(x) = 0$ when $x = -4$ and by definition $f(x) \geqslant 0$ and is symmetrical about the line $x = -4$.

We now find the value of $f(x)$ at two points either side of $x = -4$ and sketch the function $f(x)$.

When $x = -3$

$$f(x) = |-3 + 4| = |1| = 1$$

When $x = -5$

$$f(x) = |-5 + 4| = |-1| = 1$$

We are now able to sketch $f(x)$.

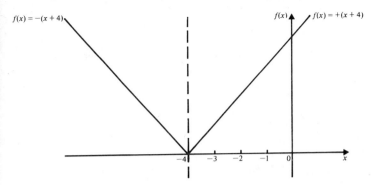

$g(x) = 0$ when $x = -3$ and by definition $g(x) \geqslant 0$ and is symmetrical about the line $x = -3$.

We now find the value of $g(x)$ at two points either side of $x = -3$ and sketch the function $g(x)$.

When $x = -2$

$$g(x) = |-2 + 3| = |1| = 1$$

When $x = -4$

$$g(x) = |-4 + 3| = |-1| = 1$$

We are now able to sketch $g(x)$.

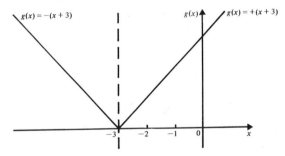

We can now sketch the two functions on the same diagram.

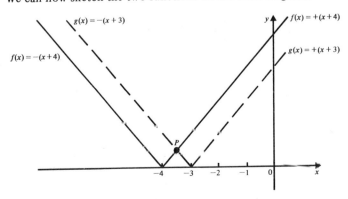

From the diagram, $f(x) < g(x)$ for all values of x to the left of the intersection point P.

3

At P the *positive branch* of $f(x)$ meets the *negative branch* of $g(x)$.

Therefore

$$(x + 4) = -(x + 3)$$

$\Rightarrow \qquad x + 4 = -x - 3$

$\Rightarrow \qquad 2x + 4 = -3$

$\Rightarrow \qquad 2x = -7$

$\Rightarrow \qquad x = -3\frac{1}{2}$

Therefore $f(x) < g(x)$ and $|x + 4| < |x + 3|$ for $\{x : x < -3\frac{1}{2}\}$.

In order to evaluate this area we sketch both graphs and find their points of intersection.

The line $y = 4x$ passes through the origin and has a gradient of 4.

The curve $y^2 = 16x$ is recognised as a parabola with symmetry about the x-axis. However we will work out some points to help with the sketch.

When

$$x = 0 \qquad y^2 = 0$$

$\Rightarrow \quad y = 0$ [1]

When

$$x = 1 \qquad y^2 = 16$$

$\Rightarrow \quad y = \pm 4$ [2]

When

$$x = 2 \qquad y^2 = 32$$

$\Rightarrow \quad y = \pm\sqrt{32} = \pm 5.7 \quad$ approximately

As y^2 is always positive x can not take negative values.

4

We can now sketch both graphs on the same diagram.

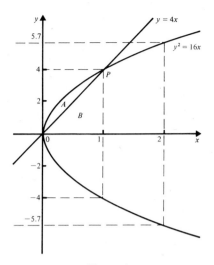

Figure 1

$y^2 = 16x \implies y = +4\sqrt{x}$ for *top* part of the curve

$y^2 = 16x \implies y = -4\sqrt{x}$ for *bottom* part of the curve

From the diagram the points of intersection are the origin and the point P.

The point P can be found by solving the equation

$$4x = +4\sqrt{x} \quad \text{(see Figure 1, above)}$$

$\implies \qquad x = \sqrt{x}$

$\implies \qquad x^2 = x$

$\implies \qquad x^2 - x = 0$

$\implies \qquad x(x-1) = 0$

$\implies \qquad$ either $\quad x = 0 \quad$ or $\quad x - 1 = 0$

$\implies \qquad\qquad x = 0 \quad$ or $\quad 1$

When

$$x = 0 \qquad y = 0 \qquad \text{from [1]}$$

When

$$x = 1 \qquad y = +4 \qquad \text{from [2]}$$

Therefore the points of intersection are the origin, $(0, 0)$, and the point $P(1, 4)$.

We find the required area, A in Figure 1, above, by evaluating the area between the curve $y = +4\sqrt{x}$, the x-axis and $x = 1$, and subtracting the area B in Figure 1.

The area between a curve, the x-axis and the lines $x = a$ and $x = b$ is given by the formula

$$\text{Area} = \int_a^b y \, dx \qquad\qquad [3]$$

5

(The formula applies only if the curve does not cross the x-axis between the limits x = a and x = b.)

Therefore the area $(A + B)$ is given by

$$A + B = \int_0^1 4\sqrt{x}\, dx \quad \text{from [3]}$$

$$\Rightarrow \quad A + B = \int_0^1 4x^{1/2}\, dx$$

$$\Rightarrow \quad A + B = \left[\tfrac{8}{3}x^{3/2}\right]_0^1$$

$$\Rightarrow \quad A + B = (\tfrac{8}{3}(1)^{3/2}) - (0)$$

$$\Rightarrow \quad A + B = \tfrac{8}{3} = 2\tfrac{2}{3} \qquad\qquad [4]$$

From Figure 1, area B is a right-angled triangle with area

$$\tfrac{1}{2} \times 1 \times 4$$

$$\Rightarrow \quad B = 2 \qquad\qquad [5]$$

Therefore the area, A , between the line $y = 4x$ and the curve $y^2 = 16x$ is

$$2\tfrac{2}{3} - 2 = \tfrac{2}{3} \quad \text{subtracting [5] from [4]}$$

4

$$\cdot R \sin(x - \alpha) \equiv R \sin x \cos \alpha - R \cos x \sin \alpha \qquad [1]$$

Putting

$$\sin x - \cos x \equiv R \sin(x - \alpha) \qquad\qquad [2]$$

$$\Rightarrow \quad \sin x - \cos x \equiv R \sin x \cos \alpha - R \cos x \sin \alpha \quad \text{from [1]}$$
$$[3]$$

For statement [3] to be true the term in sin x on the L.H.S. must be equivalent to the term in sin x on the R.H.S. Similarly the terms in cos x must be equivalent.

Therefore

$$\sin x \equiv R \sin x \cos \alpha \quad \text{from [3]}$$

$$\Rightarrow \quad R \cos \alpha = 1 \qquad\qquad [4]$$

and

$$-\cos x \equiv -R \cos x \sin \alpha \quad \text{from [3]}$$

$$\Rightarrow \quad R \sin \alpha = 1 \qquad\qquad [5]$$

We now use equations [4] and [5] to find R and α.

Dividing [5] by [4] gives

$$\frac{R \sin \alpha}{R \cos \alpha} = \frac{1}{1}$$

$$\Rightarrow \quad \tan \alpha = 1$$

$$\Rightarrow \quad \alpha = \frac{\pi}{4} \quad \text{(since α is acute)} \qquad\qquad [6]$$

Squaring and adding [5] and [4] gives

$$R^2 \sin^2\alpha + R^2 \cos^2\alpha = 1^2 + 1^2$$

$\Rightarrow \qquad R^2(\sin^2\alpha + \cos^2\alpha) = 2$

$\Rightarrow \qquad\qquad R^2 = 2 \quad \text{since} \quad \sin^2\alpha + \cos^2\alpha = 1$

$\Rightarrow \qquad\qquad R = \sqrt{2} \qquad\qquad\qquad\qquad\qquad [7]$

Therefore

$$\sin x - \cos x \equiv \sqrt{2}\sin\left(x - \frac{\pi}{4}\right) \quad \text{from [2], [6] and [7]}$$

$$[8]$$

We now use equation [8] to solve the equation

$$\sin x - \cos x = 1$$

From [8]

$$\sqrt{2}\sin\left(x - \frac{\pi}{4}\right) = 1$$

$\Rightarrow \qquad \sin\left(x - \frac{\pi}{4}\right) = \frac{1}{\sqrt{2}}$

$\Rightarrow \qquad \left(x - \frac{\pi}{4}\right) = \frac{\pi}{4} \quad \text{or} \quad \frac{3\pi}{4},$

$\Rightarrow \qquad\qquad x = \frac{\pi}{2} \quad \text{or} \quad \pi, \qquad \text{for } 0 \leqslant x \leqslant 2\pi$

$\Rightarrow \qquad\qquad x = \frac{\pi}{2} \quad \text{or} \quad \pi \quad \text{plus multiples of } 2\pi \text{ for the general}$
$\qquad\qquad\qquad\qquad\qquad\qquad \text{solution}$

$\Rightarrow \qquad\qquad x = 2n\pi + \frac{\pi}{2} \quad \text{or} \quad 2n\pi + \pi \quad \text{for } n = 0, \pm1,$
$\qquad\qquad\qquad\qquad\qquad\qquad\qquad\qquad\qquad \pm2, \pm3, \ldots$

Therefore the general solution of $\sin x - \cos x = 1$ is

$$x = 2n\pi + \frac{\pi}{2} \quad \text{or} \quad (2n+1)\pi \quad \text{for } n = 0, \pm1, \pm2, \ldots$$

5

$$\sum_{r=1}^{n+1} u_r \equiv u_1 + u_2 + u_3 + \ldots + u_n + u_{n+1} \qquad\qquad [1]$$

$$\rightarrow \quad \sum_{r=1}^{n+1} u_r \equiv u_1 + ku_1 + k^2u_1 + \ldots + k^{n-1}u_1 + k^n u_1 \qquad [2]$$
$$\text{(since } k \text{ is the common ratio)}$$

Comparing [1] and [2] gives

$$u_2 \equiv ku_1, \quad u_3 \equiv k^2u_1, \ldots, \quad u_n \equiv k^{n-1}u_1$$

and $\quad u_{n+1} \equiv k^n u_1 \qquad\qquad\qquad\qquad\qquad\qquad\qquad [3]$

7

To show that the series $\sum_{r=1}^{n} (u_r u_{r+1})$ *is a G.P. we write out its terms and find the common ratio.*

$$\sum_{r=1}^{n} (u_r u_{r+1}) \equiv u_1 u_2 + u_2 u_3 + u_3 u_4 + \ldots + u_n u_{n+1}$$

$$\Rightarrow \quad \sum_{r=1}^{n} (u_r u_{r+1}) \equiv u_1 k u_1 + k u_1 k^2 u_1 + k^2 u_1 k^3 u_1 + \ldots$$
$$+ k^{n-1} u_1 k^n u_1 \quad \text{from [3]}$$

$$\Rightarrow \quad \sum_{r=1}^{n} (u_r u_{r+1}) \equiv k u_1^2 + k^3 u_1^2 + k^5 u_1^2 + \ldots + k^{2n-1} u_1^2 \quad [4]$$

Therefore the series $\sum_{r=1}^{n} (u_r u_{r+1})$ is a geometric series with first term $k u_1^2$ and common ratio k^2.

Using the formula for the sum of a G.P.

$$\frac{a(1 - r^n)}{1 - r}$$

(where $a \equiv$ first term, $n \equiv$ number of terms and $r \equiv$ common ratio)

with $a = k u_1^2$ and $r = k^2$ from [5]

$$\sum_{r=1}^{n} (u_r u_{r+1}) = \frac{k u_1^2 [1 - (k^2)^n]}{(1 - k^2)}$$

Therefore

$$\sum_{r=1}^{n} (u_r u_{r+1}) = \frac{U_1^2 k (1 - k^{2n})}{(1 - k^2)}$$

6

To show $f(x) > 0$ we find the range of $f(x)$.

We begin by completing the square on $f(x)$.

$$f(x) \equiv x^2 - 6x + 10 \qquad [1]$$
$$\Rightarrow \qquad f(x) \equiv (x - 3)^2 + 1 \qquad [2]$$

Since

$$(x - 3)^2 \geqslant 0 \qquad [3]$$
$$(x - 3)^2 + 1 \geqslant 1$$
$$\Rightarrow \qquad f(x) \geqslant 1 \qquad [4]$$
$$\Rightarrow \qquad f(x) > 0$$

8

The graph $y = f(x)$ is recognised as a quadratic curve with the following features:

(i) a minimum value of 1   from [4];
(ii) this minimum value occurs when $x = 3$ from [3];
(iii) when $x = 0$, $y = f(x) = 10$ from [1].

This information is used to produce Figure 1.

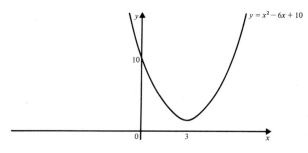

$y = x^2 - 6x + 10$

10

0   3

Figure 1

To sketch the graph of $y = \dfrac{1}{f(x)}$ we substitute $f(x)$ from [2]

to give $y = \dfrac{1}{(x-3)^2 + 1}$

When sketching curves of this type we should consider the following:

(i) *does the curve cross either axis (i.e. what happens when $x = 0$ and $y = 0$)?*

(ii) *are there any values of x for which the denominator is zero (i.e. are there any vertical asymptotes)?*

(iii) *the values of y for large values of x – both positive and negative (i.e. are there any horizontal asymptotes?).*

(iv) *how does the curve approach its asymptotes?*

In this type of question the emphasis is more on investigation than algebraic technique. It is very important that we explain clearly how the function behaves at each of the steps (i)–(iv) above. We can then mark this information on a diagram (e.g. Figure 2, overleaf) to produce a skeleton sketch. It will then become clear how the sketch may be completed (see Figure 3, overleaf).

(i) (a) When $x = 0$, $y = \dfrac{1}{(-3)^2 + 1} = \dfrac{1}{9+1} = \dfrac{1}{10}$.

(b) There is no value of x for which $y = \dfrac{1}{f(x)} = 0$.

(ii) There are no vertical asymptotes because the denominator is never zero   from [4].

(iii) and (iv) As $x \to \pm\infty$, $y \to 0$ (and is positive)

This information is used to produce Figure 2, overleaf.

9

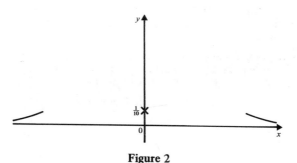

Figure 2

From [4], the minimum value of $f(x)$ is 1, hence the maximum value of $\dfrac{1}{f(x)}$ is 1. (This maximum occurs when $x = 3$.) This information, together with Figure 2, is used to produce Figure 3.

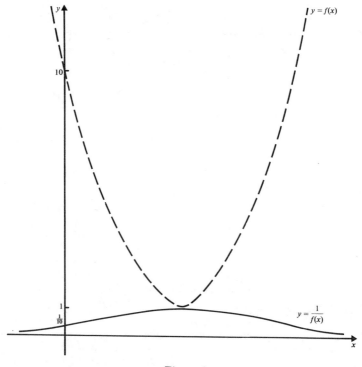

Figure 3

<hr>

7

Let $\quad f(x) \equiv \dfrac{1}{(x + 1)(x + 3)} \equiv \dfrac{A}{(x + 1)} + \dfrac{B}{(x + 3)}$ [1]

Therefore

$$\frac{1}{(x + 1)(x + 3)} \equiv \frac{A(x + 3) + B(x + 1)}{(x + 1)(x + 3)}$$

10

If we compare the numerators, then

$$1 \equiv A(x + 3) + B(x + 1) \qquad [2]$$

To find B we will make A 'disappear' by putting $x = -3$ in [2]

$$\Rightarrow \quad 1 = A(-3 + 3) + B(-3 + 1)$$

$$\Rightarrow \quad 1 = -2B$$

$$\Rightarrow \quad B = -\tfrac{1}{2}$$

To find A we will make B 'disappear' by putting $x = -1$ in [2]

$$\Rightarrow \quad 1 = A(-1 + 3) + B(-1 + 1)$$

$$\Rightarrow \quad 1 = 2A$$

$$\Rightarrow \quad A = \tfrac{1}{2}$$

Hence, substituting $A = \tfrac{1}{2}$ and $B = -\tfrac{1}{2}$ into [1] gives

$$f(x) \equiv \frac{1}{(x + 1)(x + 3)} \equiv \frac{\tfrac{1}{2}}{(x + 1)} + \frac{-\tfrac{1}{2}}{(x + 3)}$$

Therefore

$$f(x) \equiv \frac{1}{(x + 1)(x + 3)} \equiv \frac{1}{2(x + 1)} - \frac{1}{2(x + 3)} \qquad [3]$$

Using [3]

$$\frac{1}{(2r + 1)(2r + 3)} \equiv \frac{1}{2(2r + 1)} - \frac{1}{2(2r + 3)}$$

$$\Rightarrow \quad \sum_{r=1}^{n} \frac{1}{(2r + 1)(2r + 3)} \equiv \sum_{r=1}^{n} \left(\frac{1}{2(2r + 1)} - \frac{1}{2(2r + 3)} \right)$$

To sum this series we substitute $r = 1, 2, 3, \ldots, n$ and list the fractions as follows:

r	$\dfrac{1}{2(2r + 1)}$	$-\dfrac{1}{2(2r + 3)}$
1	$\tfrac{1}{6}$	$-\tfrac{1}{10}$
2	$\tfrac{1}{10}$	$-\tfrac{1}{14}$
3	$\tfrac{1}{14}$	$-\tfrac{1}{18}$
4	$\tfrac{1}{18}$	.
$\vdots$	$\vdots$	$\vdots$
$n-1$	$\dfrac{1}{2(2n-2+1)} = \dfrac{1}{2(2n-1)}$	$-\dfrac{1}{2(2n-2+3)} = -\dfrac{1}{2(2n+1)}$
n	$\dfrac{1}{2(2n+1)}$	$-\dfrac{1}{2(2n+3)}$

On summation all the fractions cancel apart from the top one in the middle column and the bottom one in the end column.

Therefore

$$\sum_{r=1}^{n} \frac{1}{(2r+1)(2r+3)} = \frac{1}{6} - \frac{1}{2(2n+3)}$$

$$= \frac{(2n+3)-3}{6(2n+3)}$$

$$= \frac{2n}{6(2n+3)}$$

Therefore

$$\sum_{r=1}^{n} \frac{1}{(2r+1)(2r+3)} = \frac{n}{3(2n+3)}$$

8

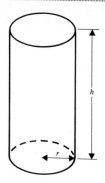

Let Figure 1 represent the cylinder with radius r m and height h m.

Figure 1

The volume, V, of the cylinder is given by

$$V = \pi r^2 h \qquad [1]$$

$$\Rightarrow \quad 16\pi = \pi r^2 h$$

$$\Rightarrow \quad 16 = r^2 h$$

$$\Rightarrow \quad h = \frac{16}{r^2} \qquad [2]$$

The exterior surface area, A, of the cylinder is given by

$$A = 2\pi r^2 + 2\pi rh \qquad [3]$$

where $2\pi r^2$ is the area of the two circular ends and $2\pi rh$ is the curved area.

In order to express A in terms of r only we must eliminate h from [3].

Substituting [2] into [3] gives

$$A = 2\pi r^2 + 2\pi r \left(\frac{16}{r^2}\right)$$

$$\Rightarrow \quad A = 2\pi r^2 + \frac{32\pi}{r}$$

The area of the cylinder is a minimum when $\dfrac{dA}{dr} = 0$

$$A = 2\pi r^2 + \frac{32\pi}{r}$$

$\Rightarrow \qquad A = 2\pi r^2 + 32\pi r^{-1}$

$\Rightarrow \qquad \dfrac{dA}{dr} = 4\pi r - 32\pi r^{-2}$ [4]

$\Rightarrow \qquad \dfrac{dA}{dr} = 4\pi r - \dfrac{32\pi}{r^2}$

$\dfrac{dA}{dr} = 0$ when $4\pi r - \dfrac{32\pi}{r^2} = 0$

$\Rightarrow \qquad 4\pi r^3 - 32\pi = 0$

$\Rightarrow \qquad r^3 - 8 = 0$

$\Rightarrow \qquad r^3 = 8$

$\Rightarrow \qquad r = 2$ [5]

To confirm this value of r produces a minimum rather than a maximum area we check the sign of $\dfrac{d^2A}{dr^2}$ where $r = 2$.

$\dfrac{dA}{dr} = 4\pi r - 32\pi r^{-2}$ equation [4]

$\Rightarrow \qquad \dfrac{d^2A}{dr^2} = 4\pi + 64\pi r^{-3}$

$\Rightarrow \qquad \dfrac{d^2A}{dr^2} = 4\pi + \dfrac{64\pi}{r^3}$

When $r = 2$

$\dfrac{d^2A}{dr^2} = 4\pi + \dfrac{64\pi}{8} = 12\pi$ which is greater than zero

Therefore with $r = 2$, A is a minimum.

Substituting [5] into [2] gives

$$h = \frac{16}{4}$$

$\Rightarrow \qquad h = 4$

Therefore the height of the cylinder is 4 m and the base radius is 2 m when the total surface area is a minimum.

9

When sketching curves of this type we should consider the following:

(i) does the curve cross either axis (i.e. what happens when $x = 0$ and $y = 0$)?

(ii) *are there any values of x for which the denominator is zero
(i.e. are there any vertical asymptotes)?*
(iii) *the values of y for large values of x — both positive and
negative (i.e. are there any horizontal asymptotes?).*
(iv) *how does the curve approach its asymptotes?*

*In this type of question the emphasis is more on investigation
than algebraic technique. It is very important that we explain
clearly how the function behaves at each of the steps (i)–(iv)
above. We can then mark this information on a diagram (e.g.
Figure 1, below) to produce a skeleton sketch. It will then
become clear how the sketch may be completed (see Figure 2,
opposite).*

$$g: x \to \frac{2x + 5}{x - 3} \quad \Rightarrow \quad g(x) = \frac{2x + 5}{x - 3}$$

For $y \equiv g(x)$

(i) (a) When $x = 0$,

$$y = -\tfrac{5}{3}$$

(b) When $y = 0$,

$$2x + 5 = 0$$

$$\Rightarrow \qquad 2x = -5$$

$$\Rightarrow \qquad x = -2\tfrac{1}{2}$$

(ii) $x - 3 = 0$ when $x = 3$

$$\Rightarrow \qquad x = 3 \quad \text{is a vertical asymptote}$$

(iii) (a) As $x \to +\infty$, $y \to 2$ (and is greater than 2).

(b) As $x \to -\infty$, $y \to 2$ (and is less than 2).

(iv) We now check values of x close to either side of the vertical
asymptote, say $x = 3.1$ and $x = 2.9$.

When $x = 3.1$,

$$y = \frac{6.2 + 5}{3.1 - 3} = \frac{11.2}{0.1} = 112 \quad \text{(i.e. large and positive)}$$

When $x = 2.9$,

$$y = \frac{5.8 + 5}{2.9 - 3} = \frac{10.8}{-0.1} = -108 \quad \text{(i.e. large and negative)}$$

This information is used to produce Figure 1.

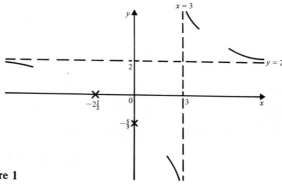

Figure 1

14

The information in Figure 1, opposite, implies that the graph of $g(x) = \dfrac{2x + 5}{x - 3}$ is as shown in Figure 2.

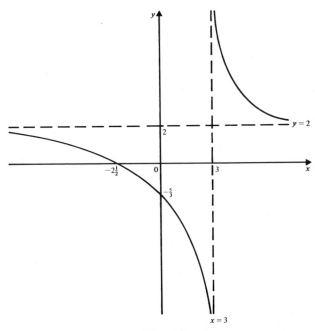

Figure 2

To find the inverse of a function g(x) we form the equation
y = g(x) and rearrange it to make x the subject of the formula
i.e. x = g⁻¹(y). This formula when written in terms of x, not y,
will be the inverse function g⁻¹(x).

$$g(x) = \frac{2x + 5}{x - 3}$$

$$\Rightarrow \quad y = \frac{2x + 5}{x - 3}$$

$$\Rightarrow \quad y(x - 3) = 2x + 5$$

$$\Rightarrow \quad yx - 3y = 2x + 5$$

$$\Rightarrow \quad yx = 2x + 5 + 3y$$

$$\Rightarrow \quad yx - 2x = 5 + 3y$$

$$\Rightarrow \quad x(y - 2) = 5 + 3y$$

$$\Rightarrow \quad x = \frac{5 + 3y}{y - 2}$$

$$\Rightarrow \quad g^{-1}(y) = \frac{3y + 5}{y - 2}$$

$$\Rightarrow \quad g^{-1}(x) = \frac{3x + 5}{x - 2}$$

15

The range of a function g(x) is the domain of its inverse $g^{-1}(x)$.

From Figure 2, $g(x)$ takes all real values except 2.

Therefore the range of $g(x)$ is

$$x \in \mathbb{R}, \qquad x \neq 2$$

Therefore the domain of $g^{-1}(x)$ is also

$$x \in \mathbb{R}, \qquad x \neq 2$$

Therefore $\qquad g^{-1}(x) = \dfrac{3x + 5}{x - 2}, \quad x \in \mathbb{R}, \quad x \neq 2$

10

*If these two lines intersect there will exist unique values of s
and t for which the lines have a common point.*

When $s = 0$ and $t = 0$

$\qquad \mathbf{r} = \mathbf{k}$ for both lines

Therefore the position vector of A, $\mathbf{r}_A$, is given by

$$\mathbf{r}_A = \mathbf{k} \tag{1}$$

*If the vector $\mathbf{p}$ is perpendicular to the line $r = a + \lambda b$ then the
scalar product of $\mathbf{p}$ and $\mathbf{b}$ (i.e. the 'direction' of the line) is zero.*

The direction vectors of the two lines are $(\mathbf{i} + \mathbf{j})$ and $(-\mathbf{i} + \mathbf{k})$.

Let $x\mathbf{i} + y\mathbf{j} + z\mathbf{k}$ be a vector perpendicular to both $(\mathbf{i} + \mathbf{j})$ and
$(-\mathbf{i} + \mathbf{k})$.

Therefore

$$(\mathbf{i} + \mathbf{j}).(x\mathbf{i} + y\mathbf{j} + z\mathbf{k}) = 0 \tag{2}$$

and $\quad (-\mathbf{i} + \mathbf{k}).(x\mathbf{i} + y\mathbf{j} + z\mathbf{k}) = 0 \tag{3}$

From [2]

$$(\mathbf{i} + \mathbf{j}).(x\mathbf{i} + y\mathbf{j} + z\mathbf{k}) = 0$$

$\Rightarrow \qquad\qquad\qquad x + y = 0$

$\Rightarrow \qquad\qquad\qquad\qquad y = -x \tag{4}$

From [3]

$$(-\mathbf{i} + \mathbf{k}).(x\mathbf{i} + y\mathbf{j} + z\mathbf{k}) = 0$$

$\Rightarrow \qquad\qquad\qquad -x + z = 0$

$\Rightarrow \qquad\qquad\qquad\qquad z = x \tag{5}$

Therefore

$$x\mathbf{i} + y\mathbf{j} + z\mathbf{k} \equiv x\mathbf{i} - x\mathbf{j} + x\mathbf{k} \quad \text{from [4] and [5]}$$
$$\equiv x(\mathbf{i} - \mathbf{j} + \mathbf{k})$$

Therefore a vector perpendicular to both lines is

$$\mathbf{i} - \mathbf{j} + \mathbf{k} \tag{6}$$

The general vector equation of a plane is $\mathbf{r} \cdot \mathbf{n} = d$ where $\mathbf{r}$ represents a general position vector in the plane; $\mathbf{n}$ is a normal vector to the plane, and d is a scalar.

d can be evaluated if $\mathbf{n}$ and a particular value of $\mathbf{r}$ are known.

$$\mathbf{n} = \mathbf{i} - \mathbf{j} + \mathbf{k} \quad \text{from [6]} \qquad [7]$$

A particular value of $\mathbf{r}$ is

$$\mathbf{r} = \mathbf{k} \quad \text{from [1]}$$

Therefore

$$\mathbf{r} \cdot \mathbf{n} = d$$

$\Rightarrow \qquad \mathbf{k} \cdot (\mathbf{i} - \mathbf{j} + \mathbf{k}) = d$

$\Rightarrow \qquad\qquad 1 = d \qquad [8]$

Therefore the vector equation of the plane is

$$\mathbf{r} \cdot (\mathbf{i} - \mathbf{j} + \mathbf{k}) = 1 \quad \text{from [7] and [8]}$$

11.

$f(x) \equiv x^3 + 2x + 4$

$\Rightarrow \qquad f'(x) \equiv 3x^2 + 2 \qquad\qquad\qquad [1]$

$\Rightarrow \qquad f'(x)$ is always positive

$\Rightarrow \qquad$ the gradient of $y = f(x)$ is always positive

$\qquad\qquad f'(x) = 0$

$\Rightarrow \qquad 3x^2 + 2 = 0 \quad \text{from [1]}$

$\Rightarrow \qquad\quad 3x^2 = -2 \qquad\qquad\qquad [2]$

Equation [2] has no real roots

$\Rightarrow \qquad f'(x) = 0 \quad$ has no real roots

$\Rightarrow \qquad\quad y = f(x) \quad$ has no turning points

As the graph of $y = f(x)$ has no turning points and has always a positive gradient it can only cross the x-axis on one occasion.

Therefore $f(x) = 0$ has only one real root.

As any curve, f(x), crosses the x-axis the value of f(x) will change from positive to negative, or vice versa.

$$f(-2) = -8 - 4 + 4 = -8$$
$$f(-1) = -1 - 2 + 4 = 1$$

From the above information we see that $f(x)$ changes sign between

$$x = -2 \quad \text{and} \quad x = -1$$

Therefore there must be a value of x for which $f(x) = 0$ in the interval

$$-2 < x < -1$$

$$x_1 = -1 \qquad\qquad\qquad [3]$$

$$x_{n+1} = -\tfrac{1}{6}(x_n^3 - 4x_n + 4) \qquad\qquad\qquad [4]$$

17

When using an iterative procedure to find the root of an equation we substitute integer values for n (starting with $n = 1$).

Substituting $n = 1$ into [4]

$$x_2 = -\tfrac{1}{6}(x_1{}^3 - 4x_1 + 4)$$

$\Rightarrow \quad x_2 = -\tfrac{1}{6}[(-1)^3 - 4(-1) + 4] \quad$ from [3]

$\Rightarrow \quad x_2 = -\tfrac{1}{6}(-1 + 4 + 4)$

$\Rightarrow \quad x_2 = -\tfrac{1}{6}(7)$

$\Rightarrow \quad x_2 = -\tfrac{7}{6} \quad$ is a second approximation $\hfill$ [5]

Substituting $n = 2$ into [4]

$$x_3 = -\tfrac{1}{6}(x_2{}^3 - 4x_2 + 4)$$

$\Rightarrow \quad x_3 = -\tfrac{1}{6}[(-\tfrac{7}{6})^3 - 4(-\tfrac{7}{6}) + 4] \quad$ from [5]

$\Rightarrow \quad x_3 = -\tfrac{1}{6}[-1.5880 + 4.6667 + 4]$

$\Rightarrow \quad x_3 = -\tfrac{1}{6}(7.0787)$

$\Rightarrow \quad x_3 = -1.1798$

$\Rightarrow \quad x_3 = -1.18 \quad$ to 2 decimal places

Therefore the third (and final) approximation is

$\qquad -1.18 \quad$ to 2 decimal places

12

$|x| < 1$ is stated to ensure that the Binomial Theorem will give a convergent series in ascending powers of x. Apart from this $|x| < 1$ has no relevance within the question.

$$\frac{1 - kx^2}{(1 - x^2)^{1/2}} \equiv (1 - kx^2)(1 - x^2)^{-1/2}$$

We now expand $(1 - x^2)^{-1/2}$ using the Binomial Theorem

$$(1 + y)^n = 1 + ny + \frac{n(n - 1)}{2!}y^2$$

$$+ \frac{n(n - 1)(n - 2)}{3!}y^3 + \dots$$

with $\qquad y = -x^2 \quad$ and $\quad n = -\tfrac{1}{2}$

Therefore

$$(1 - x^2)^{-1/2} = 1 + (-\tfrac{1}{2})(-x^2) + \frac{(-\tfrac{1}{2})(-\tfrac{3}{2})(-x^2)^2}{2!}$$

$$+ \frac{(-\tfrac{1}{2})(-\tfrac{3}{2})(-\tfrac{5}{2})(-x^2)^3}{3!} + \dots$$

$\Rightarrow \quad (1 - x^2)^{-1/2} = 1 + \tfrac{1}{2}x^2 + \tfrac{3}{8}x^4 + \tfrac{5}{16}x^6 + \dots$

Therefore

$$\frac{(1 - kx^2)}{(1 - x^2)^{1/2}} \equiv (1 - kx^2)(1 + \tfrac{1}{2}x^2 + \tfrac{3}{8}x^4 + \tfrac{5}{16}x^6 + \dots)$$

$\Rightarrow \qquad \dfrac{1-kx^2}{(1-x^2)^{1/2}} = 1 + \frac{1}{2}x^2 + \frac{3}{8}x^4 + \frac{5}{16}x^6$

$$+ \ldots - kx^2 - \frac{1}{2}kx^4 - \frac{3}{8}kx^6 - \frac{5}{16}kx^8 - \ldots$$

$$= 1 + (\tfrac{1}{2}-k)x^2 + (\tfrac{3}{8}-\tfrac{1}{2}k)x^4 + (\tfrac{5}{16}-\tfrac{3}{8}k)x^6 + \ldots$$

If the coefficient of x^2 is given as zero

$$(\tfrac{1}{2}-k) = 0$$

$\Rightarrow \qquad k = \frac{1}{2}$

For this value of k

$$\dfrac{1-kx^2}{(1-x^2)^{1/2}} = 1 + (\tfrac{1}{2}-\tfrac{1}{2})x^2 + (\tfrac{3}{8}-\tfrac{1}{4})x^4 + (\tfrac{5}{16}-\tfrac{3}{16})x^6 + \ldots$$

Therefore the first three non-zero terms in the expansion of

$$\dfrac{1-kx^2}{(1-x^2)^{1/2}} \quad \text{are} \quad 1 + \tfrac{1}{8}x^4 + \tfrac{1}{8}x^6 \qquad \text{when} \quad k = \tfrac{1}{2}$$

13

To solve the differential equation we will separate dx from dy and collect terms in y and dy on the L.H.S. and terms in x and dx on the R.H.S.

$$\frac{dy}{dx} = \frac{(y^2-1)}{x}$$

$\Rightarrow \qquad dy = \dfrac{(y^2-1)}{x}\, dx$

$\Rightarrow \qquad \dfrac{dy}{(y^2-1)} = \dfrac{dx}{x}$

$\Rightarrow \qquad \displaystyle\int \dfrac{dy}{(y^2-1)} = \int \dfrac{dx}{x} \qquad\qquad [1]$

To integrate the L.H.S. we express $\dfrac{1}{(y^2-1)}$ *in partial fractions.*

Let $\qquad \dfrac{1}{y^2-1} \equiv \dfrac{1}{(y+1)(y-1)} \equiv \dfrac{A}{(y+1)} + \dfrac{B}{(y-1)}$

$\qquad\qquad\qquad\qquad\qquad\qquad\qquad\qquad\qquad [2]$

Therefore $\qquad \dfrac{1}{(y+1)(y-1)} \equiv \dfrac{A(y-1)+B(y+1)}{(y+1)(y-1)}$

If we compare the numerators, then

$$1 \equiv A(y-1) + B(y+1) \qquad\qquad [3]$$

To find B we will make A 'disappear' by putting

$\qquad y = 1 \quad \text{in [3]}$

$\Rightarrow \qquad 1 = A(1-1) + B(1+1)$

$\Rightarrow \qquad 1 = 2B$

$\Rightarrow \qquad B = \frac{1}{2}$

To find A we will make B 'disappear' by putting $y = -1$ in [3]

$\Rightarrow \qquad 1 = A(-1 - 1) + B(-1 + 1)$

$\Rightarrow \qquad 1 = -2A$

$\Rightarrow \qquad A = -\frac{1}{2}$

Substituting $A = -\frac{1}{2}$ and $B = \frac{1}{2}$ into [2] gives

$$\frac{1}{(y^2 - 1)} = \frac{-\frac{1}{2}}{(y + 1)} + \frac{\frac{1}{2}}{(y - 1)} \qquad\qquad [4]$$

Substituting [4] into [1] gives

$$\int \left(\frac{\frac{1}{2}}{y - 1} - \frac{\frac{1}{2}}{y + 1} \right) dy = \int \frac{dx}{x}$$

$\Rightarrow \qquad \dfrac{1}{2} \displaystyle\int \dfrac{dy}{y - 1} - \dfrac{1}{2} \int \dfrac{dy}{y + 1} = \int \dfrac{dx}{x}$

$\Rightarrow \qquad \frac{1}{2} \ln (y - 1) - \frac{1}{2} \ln (y + 1) = \ln x + \ln c$

In integration involving logarithms the constant of integration may also be taken as a logarithm.

$\Rightarrow \qquad \frac{1}{2}[\ln (y - 1) - \ln (y + 1)] = \ln cx \qquad$ by one of the laws of logarithms

$\Rightarrow \qquad \frac{1}{2} \ln \left(\dfrac{y - 1}{y + 1} \right) = \ln cx \qquad$ by one of the laws of logarithms

$\Rightarrow \qquad \ln \left(\dfrac{y - 1}{y + 1} \right)^{1/2} = \ln cx \qquad$ by one of the laws of logarithms

$\Rightarrow \qquad \left(\dfrac{y - 1}{y + 1} \right)^{1/2} = cx$

$\Rightarrow \qquad \left(\dfrac{y - 1}{y + 1} \right) = c^2 x^2$

$\Rightarrow \qquad \left(\dfrac{y - 1}{y + 1} \right) = kx^2 \quad$ (where $k = c^2$) $\quad$ [5]

To find the value of k we substitute the given information $y = 2$ when $x = 1$.

$\Rightarrow \qquad \left(\dfrac{2 - 1}{2 + 1} \right) = k(1)^2$

$\Rightarrow \qquad \frac{1}{3} = k$

Substituting $k = \frac{1}{3}$ into [5] gives

$$\frac{y - 1}{y + 1} = \tfrac{1}{3} x^2$$

$\Rightarrow \qquad y - 1 = \tfrac{1}{3} x^2 (y + 1)$

$\Rightarrow \qquad y - 1 = \tfrac{1}{3} x^2 y + \tfrac{1}{3} x^2$

$$\Rightarrow \qquad\qquad 3y - 3 = x^2 y + x^2$$

$$\Rightarrow \qquad\qquad\qquad 3y = x^2 y + x^2 + 3$$

$$\Rightarrow \qquad\qquad 3y - x^2 y = x^2 + 3$$

$$\Rightarrow \qquad\qquad y(3 - x^2) = x^2 + 3$$

Therefore $y = \dfrac{x^2 + 3}{3 - x^2}$ is the solution to the differential equation.

We can represent a complex number of the form $z = x + iy$ on an Argand Diagram as follows:

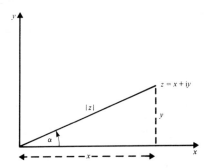

The modulus of z, $|z|$ is represented by the distance from the origin to the point z. The argument of z, arg(z), is the angle α, where $-\pi \leqslant \alpha \leqslant \pi$.

To change z_1 into the form $x + iy$ we multiply the numerator and denominator of z_1 by the complex conjugate of the denominator. This eliminates i from the denominator.

Therefore

$$z_1 = \frac{1 + i}{1 - i} = \frac{(1 + i)(1 + i)}{(1 - i)(1 + i)} = \frac{1 + i + i + i^2}{1 + i - i - i^2}$$

$$= \frac{1 + 2i - 1}{1 - -1} = \frac{2i}{2}$$

$$\Rightarrow z_1 = i \qquad\qquad\qquad\qquad [1]$$

$$|z_1| = 1$$

$$\arg(z_1) = \pi/2$$

To change z_2 into the form $x + iy$ we proceed as before:

$$z_2 = \frac{\sqrt{2}}{1-i} = \frac{\sqrt{2}(1+i)}{(1-i)(1+i)} = \frac{\sqrt{2}+\sqrt{2}i}{2}$$

(since the denominator is the same as in z_1)

$$\Rightarrow \quad z_2 = \frac{\sqrt{2}}{2} + \frac{\sqrt{2}}{2}i \qquad\qquad [2]$$

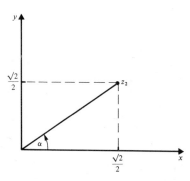

$$|z_2| = \sqrt{\left(\frac{\sqrt{2}}{2}\right)^2 + \left(\frac{\sqrt{2}}{2}\right)^2} = \sqrt{\frac{2}{4} + \frac{2}{4}} = 1$$

$$\arg(z_2) = \alpha \quad \text{where} \quad \tan\alpha = \frac{\sqrt{2}/2}{\sqrt{2}/2} = 1$$

$$\Rightarrow \quad \arg(z_2) = \pi/4$$

$$z_1 + z_2 = i + \frac{\sqrt{2}}{2} + \frac{\sqrt{2}}{2}i \quad \text{from [1] and [2]}$$

$$\Rightarrow \quad z_1 + z_2 = \frac{\sqrt{2}}{2} + \left(1 + \frac{\sqrt{2}}{2}\right)i$$

Representing z_1, z_2 and $(z_1 + z_2)$ on the same Argand Diagram gives Figure 1, below.

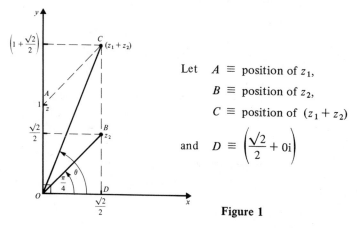

Let $A \equiv$ position of z_1,

$B \equiv$ position of z_2,

$C \equiv$ position of $(z_1 + z_2)$

and $D \equiv \left(\frac{\sqrt{2}}{2} + 0i\right)$

Figure 1

22

To deduce $\tan\frac{3}{8}\pi = 1 + \sqrt{2}$ *we consider the geometry of OABC in Figure 1, opposite.*

$OA = OB = 1$ (since $|z_1| = |z_2| = 1$)
and BC is parallel and equal to OA.

Therefore $OACB$ is a rhombus and the diagonal OC bisects angle AOB.

$$A\widehat{O}B = \pi/2 - \pi/4$$

$$= \pi/4$$

$$\Rightarrow \quad C\widehat{O}B = \pi/8$$

$$\Rightarrow \quad C\widehat{O}D = \pi/8 + \pi/4$$

$$= 3\pi/8$$

Therefore

$$\theta = 3\pi/8 \qquad\qquad [3]$$

From Figure 1, opposite

$$\tan\theta = \frac{1 + \dfrac{\sqrt{2}}{2}}{\dfrac{\sqrt{2}}{2}}$$

$$\Rightarrow \quad \tan\theta = \frac{2 + \sqrt{2}}{\sqrt{2}}$$

$$\Rightarrow \quad \tan\theta = \frac{2}{\sqrt{2}} + \frac{\sqrt{2}}{\sqrt{2}}$$

$$\Rightarrow \quad \tan\theta = \sqrt{2} + 1$$

Therefore

$$\tan\frac{3}{8}\pi = 1 + \sqrt{2} \quad \text{from } [3]$$

15

Let $(at_1{}^2, 2at_1)$ be the point, P, on the curve with parameter t_1. The gradient of the tangent to the curve at a general point $(at^2, 2at)$ is given by $\dfrac{dy}{dx}$ where

$$\frac{dy}{dx} = \frac{dy}{dt} \times \frac{dt}{dx} \qquad\qquad [1]$$

$$x = at^2$$

$\Rightarrow \quad \dfrac{dx}{dt} = 2at$

$\Rightarrow \quad \dfrac{dt}{dx} = \dfrac{1}{2at}$ [2]

$y = 2at$

$\Rightarrow \quad \dfrac{dy}{dt} = 2a$ [3]

Substituting [2] and [3] into [1] gives

$$\dfrac{dy}{dx} = 2a \times \dfrac{1}{2at} = \dfrac{1}{t}$$

$\Rightarrow \quad \dfrac{dy}{dx} = \dfrac{1}{t_1}$ is the gradient of the tangent at P [4]

The equation of a line of gradient m which passes through the point (a, b) is

$$y - b = m(x - a)$$ [5]

Therefore the equation of the tangent passing through $(at_1{}^2, 2at_1)$ with gradient $\dfrac{1}{t_1}$ is

$$y - 2at_1 = \dfrac{1}{t_1}(x - at_1{}^2) \quad \text{from [5]}$$

$\Rightarrow \quad t_1 y - 2at_1{}^2 = x - at_1{}^2$

Therefore

$$t_1 y = x + at_1{}^2 \quad \text{is the equation of the tangent at } P \quad [6]$$

When this tangent crosses the x-axis, $y = 0$. Therefore, substituting $y = 0$ into [6] gives

$$0 = x + at_1{}^2$$

$\Rightarrow \quad x = -at_1{}^2$

Therefore the point, T, where this tangent crosses the x-axis is $(-at_1{}^2, 0)$.

The gradient of the normal to a curve is

$$\dfrac{-1}{\textit{the gradient of the tangent at that point}}$$

Therefore the gradient of the normal at P is

$$\dfrac{-1}{1/t_1} = -t_1 \quad \text{from [4]}$$

Therefore the equation of the normal passing through $(at_1{}^2, 2at_1)$ with gradient $-t_1$ is

$$y - 2at_1 = -t_1(x - at_1{}^2) \quad \text{from [5]}$$

$\Rightarrow \quad y - 2at_1 = -t_1 x + at_1{}^3$

Therefore

$$y = -t_1x + at_1^3 + 2at_1 \qquad [7]$$

is the equation of the normal at P.

When this normal crosses the x-axis, $y = 0$. Therefore substituting $y = 0$ into [7] gives

$$0 = -t_1x + at_1^3 + 2at_1$$

$$\Rightarrow \quad t_1x = at_1^3 + 2at_1$$

$$\Rightarrow \quad x = at_1^2 + 2a$$

Therefore the point, N, where this normal crosses the x-axis is $(at_1^2 + 2a, 0)$.

To find PT and PN we use Pythagoras' Theorem. The distance between the two points (x_1, y_1) and (x_2, y_2) is

$$\sqrt{(x_1 - x_2)^2 + (y_1 - y_2)^2} \qquad [8]$$

When P is $(at_1^2, 2at_1)$ and T is $(-at_1^2, 0)$

$$PT = \sqrt{(at_1^2 - \overline{}at_1^2)^2 + (2at_1 - 0)^2} \quad \text{from [8]}$$

$$= \sqrt{(2at_1^2)^2 + 4a^2t_1^2}$$

$$= \sqrt{4a^2t_1^4 + 4a^2t_1^2}$$

$$= \sqrt{4a^2t_1^2(t_1^2 + 1)}$$

$$\Rightarrow \quad PT = 2a|t_1|\sqrt{t_1^2 + 1} \quad \text{(since } a \text{ is positive)} \qquad [9]$$

When P is $(at_1^2, 2at_1)$ and N is $(at_1^2 + 2a, 0)$

$$PN = \sqrt{[at_1^2 - (at_1^2 + 2a)]^2 + (2at_1 - 0)^2} \quad \text{from [8]}$$

$$= \sqrt{(-2a)^2 + 4a^2t_1^2}$$

$$= \sqrt{4a^2 + 4a^2t_1^2}$$

$$= \sqrt{4a^2(1 + t_1^2)}$$

$$\Rightarrow \quad PN = 2a\sqrt{1 + t_1^2} \qquad [10]$$

Therefore

$$\frac{PT}{PN} = \frac{2a|t_1|\sqrt{t_1^2 + 1}}{2a\sqrt{1 + t_1^2}} \quad \text{from [9] and [10]}$$

Therefore

$$\frac{PT}{PN} = |t_1|$$

June 1983

PAPER 2

(a) If we are to choose a committee with *exactly* one woman it will also include *exactly* three men.

We can choose one woman from a group of four in $^4C_1$ ways. We can choose three men from a group of eight in $^8C_3$ ways. Therefore the number of ways of selecting a committee of one woman and three men is

$$^4C_1 \times {}^8C_3 = \frac{4!}{3! \times 1!} \times \frac{8!}{5! \times 3!} = 4 \times 56 = 224$$

(b) We can calculate the number of ways of choosing a committee with *at least* one woman by calculating the number of ways a committee of any four people can be selected ($^{12}C_4$), and subtracting the number of ways a committee without women can be selected (i.e. a committee of men only, which is $^8C_4$).

Therefore the number of possible selections is

$$^{12}C_4 - {}^8C_4 = \frac{12!}{8! \times 4!} - \frac{8!}{4! \times 4!} = 495 - 70 = 425$$

2

When considering the sum and product of the roots, α and β, of a quadratic equation we write the equation in the general form of

$$ax^2 + bx + c = 0$$

$$\Rightarrow \quad x^2 + \frac{b}{a}x + \frac{c}{a} = 0 \tag{1}$$

In this case, the sum of the roots

$$\alpha + \beta = -\frac{b}{a} \tag{2}$$

and the product of the roots

$$\alpha\beta = \frac{c}{a} \tag{3}$$

Rearranging $(x - b_1)(x - b_2) = c$ into the general form, [1], gives

$$x^2 - b_1 x - b_2 x + b_1 b_2 = c$$

$$\Rightarrow \quad x^2 - x(b_1 + b_2) + b_1 b_2 = c$$

$$\Rightarrow \quad x^2 - x(b_1 + b_2) + b_1 b_2 - c = 0$$

This equation has roots a_1 and a_2

$$\Rightarrow \quad a_1 + a_2 = \frac{b_1 + b_2}{1} = b_1 + b_2 \quad \text{from [2]} \tag{4}$$

and

$$a_1 a_2 = \frac{b_1 b_2 - c}{1} = b_1 b_2 - c \quad \text{from [3]} \tag{5}$$

If $-a_1$ and $-a_2$ are the roots of a quadratic equation then the sum of the roots is

$$-a_1 + -a_2 = -(a_1 + a_2) = -(b_1 + b_2) \quad \text{from [4]}$$

and the product of the roots is

$$(-a_1)(-a_2) = a_1 a_2 = b_1 b_2 - c \quad \text{from [5]}$$

Hence the equation with roots $-a_1$ and $-a_2$ is

$$x^2 + (b_1 + b_2)x + b_1 b_2 - c = 0 \quad \text{from [1]}$$

$\Rightarrow \qquad x^2 + (b_1 + b_2)x + b_1 b_2 = c$

$\Rightarrow \qquad x^2 + b_1 x + b_2 x + b_1 b_2 = c$

$\Rightarrow \qquad (x + b_1)(x + b_2) = c$

3

To solve $y = 6^x$ for x we make x the subject of the formula.

Equations of this form can be solved by taking logarithms of both sides (either natural or base ten). This gives

$$\log y = \log 6^x$$

$\Rightarrow \qquad \log y = x \log 6 \quad \text{(by one of the laws of logarithms)}$

$\Rightarrow \qquad \dfrac{\log y}{\log 6} = x$

Therefore, when $y = 0.5$

$$x = \frac{\log 0.5}{\log 6} = -0.39 \quad \text{to 2 decimal places}$$

4

The curve is given in a parametric form. We need to investigate x and y for all values of t.

$$x = at^2 \qquad \text{and} \qquad y = at^3$$

When $t = 0 \qquad x = y = 0$

Therefore the curve passes through the origin.

When $t = 1 \qquad x = y = a$

For values of $t > 1$ both x and y are positive (since $a > 0$) and the y-value is greater than the x-value (since x is a function of t^2 and y is a function of t^3). This information is used to produce the sketch overleaf.

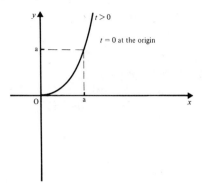

For corresponding negative values of t the x-values are the same as for $t > 0$ and the y-values will be negative.

This gives symmetry about the x-axis.

Therefore the full sketch is:

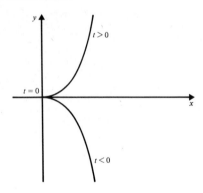

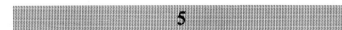

5

We could consider this question as being of the type $a \sin x + b \cos x = c$.

However we can solve most trigonometric equations involving double angles (i.e. $\sin 2x$ or $\cos 2x$) by use of the double-angle formulae. Normally they reduce to a quadratic equation.

Using $\cos 2x \equiv 2\cos^2 x - 1$ and $\sin 2x \equiv 2 \sin x \cos x$ the equation becomes

$$2\cos^2 x - 1 + 1 = 2 \sin x \cos x$$

$\Rightarrow \quad 2\cos^2 x - 2 \sin x \cos x = 0 \hfill [1]$

$\Rightarrow \quad 2 \cos x(\cos x - \sin x) = 0$

$\Rightarrow \quad$ either $\quad 2 \cos x = 0 \quad$ or $\quad \cos x - \sin x = 0$

If $\quad 2 \cos x = 0$

$\quad\quad \cos x = 0$

$$\Rightarrow \quad x = \frac{\pi}{2} \quad \text{or} \quad \frac{3}{2}\pi$$

$$\Rightarrow \quad x = \pm\frac{\pi}{2}$$

$$\Rightarrow \quad x = \pm\frac{\pi}{2} \quad \text{plus multiples of } 2\pi \text{ for the general solution}$$

$$\Rightarrow \quad x = 2n\pi \pm \frac{\pi}{2} \qquad \text{for} \quad n = 0, \pm 1, \pm 2, \ldots$$

If $\quad \cos x - \sin x = 0$

$$\cos x = \sin x$$

$$\Rightarrow \quad 1 = \frac{\sin x}{\cos x}$$

$$\Rightarrow \quad 1 = \tan x$$

$$\Rightarrow \quad x = \frac{\pi}{4} \quad \text{or} \quad \frac{5}{4}\pi \qquad \text{for} \quad 0 \leqslant x \leqslant 2\pi$$

$$\Rightarrow \quad x = \frac{\pi}{4} \quad \text{or} \quad \frac{5\pi}{4} \qquad \text{plus multiples of } 2\pi \text{ for} \\ \text{the general solution}$$

As these solutions, and all subsequent solutions, are π^c (π radians) apart this general solution can be given as

$$x = n\pi + \frac{\pi}{4} \qquad \text{for} \quad n = 0, \pm 1, \pm 2, \ldots$$

Therefore the general solution is

$$x = 2n\pi \pm \frac{\pi}{2} \quad \text{and} \quad n\pi + \frac{\pi}{4} \qquad \text{for} \quad n = 0, \pm 1, \pm 2, \ldots$$

N.B. *At [1] we must not divide the equation by $2\cos x$ or we shall lose half the solutions.*

$|6x| < 1$ *is stated to ensure that the Binomial Theorem will give a convergent series in ascending powers of x. Apart from this $|6x| < 1$ has no relevance within the question.*

Using the Binomial Theorem

$$(1 + y)^n = 1 + ny + \frac{n(n-1)}{2!} y^2 + \ldots$$

with $y = (6x)$ and $n = \frac{1}{3}$ we have

$$(1 + 6x)^{1/3} = 1 + \tfrac{1}{3}(6x) + \frac{\tfrac{1}{3}(-\tfrac{2}{3})}{2!}(6x)^2 + \ldots$$

$$= 1 + 2x - 4x^2 + \ldots$$

$$\frac{(1 + 4x)}{(1 + 2x)} \equiv (1 + 4x)(1 + 2x)^{-1}$$

We now expand $(1 + 2x)^{-1}$ using the Binomial Theorem with $y = (2x)$ and $n = -1$

$$(1 + 2x)^{-1} = 1 + (-1)(2x) + \frac{(-1)(-2)}{2!}(2x)^2 + \ldots$$

$$= 1 - 2x + 4x^2 + \ldots$$

Therefore

$$\frac{(1 + 4x)}{(1 + 2x)} \equiv (1 + 4x)(1 - 2x + 4x^2 + \ldots)$$

$$\Rightarrow \quad \frac{(1 + 4x)}{(1 + 2x)} = 1 - 2x + 4x^2 + \ldots + 4x - 8x^2 + 16x^3 + \ldots$$

$$= 1 + 2x - 4x^2 + \ldots$$

Hence the first three terms of $(1 + 6x)^{1/3}$ are identical to the first three terms of $\dfrac{(1 + 4x)}{(1 + 2x)}$.

7

When we are considering the range of a function it is sometimes useful to visualise a sketch.

As $f(x) = 2 + x - x^2$ is a 'negative' quadratic it must have a maximum point. This will happen when $f'(x) = 0$.

When we find this maximum value we shall have found the range, because $f(x)$ will be all numbers less than this value.

$$f'(x) = 1 - 2x$$

$$\Rightarrow \quad 1 - 2x = 0$$

$$\Rightarrow \quad 1 = 2x$$

$$\Rightarrow \quad \tfrac{1}{2} = x$$

If $x = \tfrac{1}{2}$, $f(x) = 2 + \tfrac{1}{2} - \tfrac{1}{4} = 2\tfrac{1}{4}$.

Therefore

$$f(x) \leqslant 2\tfrac{1}{4}$$

For a mapping to be called a function it must be either one $\rightarrow$ one or many $\rightarrow$ one. As $f(0) = 2$ and $f(1) = 2$, $f^{-1}(2)$ could be either 0 or 1. Therefore the inverse mapping, $f^{-1}(x)$, is one $\rightarrow$ many and hence not a function.

When we are considering the range of trigonometric functions it is not always easy to visualise a sketch. The following method is more suitable.

For $0 \leqslant x < \dfrac{\pi}{2}$

$$\tan x \geqslant 0$$

$$\Rightarrow \quad (1 + \tan x) \geqslant 1$$

$$\Rightarrow \quad \left(\frac{1}{1 + \tan x}\right) \leqslant 1 \quad \text{and always positive}$$

32

$\Rightarrow$ the range of $g(x)$ is $0 < g(x) \leqslant 1$

For each value of x in the domain $0 \leqslant x < \dfrac{\pi}{2}$ there is one

unique value of $\tan x$, and hence one unique value of $g(x)$.

Therefore $g(x)$ is one $\rightarrow$ one and its inverse mapping, $g^{-1}(x)$, is one $\rightarrow$ one. By definition a one $\rightarrow$ one mapping is a function. Therefore $g^{-1}(x)$ is a function.

Usually we can reduce questions on investment and compound interest to a geometric progression. The problem is to find this progression.

The initial investment is £100.

After one year it is worth £100 + 10% of £100, i.e. 110% of £100. We can write this as $(1.1)(100)$ pounds.

After a second investment of £100 he has

$\qquad (1.1)(100) + (100)$ pounds

Again, at the end of the second year he will have

$\qquad 110\%$ of $[(1.1)(100) + (100)]$ pounds

$\Rightarrow \quad (1.1)[(1.1)(100) + (100)]$ pounds

$\equiv \quad (1.1)^2(100) + (1.1)(100)$ pounds

Continuing in this manner for 19 years he will have

$\qquad (1.1)^{19}(100) + (1.1)^{18}(100) + \ldots + (1.1)^2(100)$
$\qquad + (1.1)(100)$ pounds

He then makes his twentieth payment of £100 giving him a total of

$\qquad (1.1)^{19}(100) + (1.1)^{18}(100) + \ldots + (1.1)^2(100)$
$\qquad + (1.1)(100) + (100)$ pounds

$\equiv \quad (100) + (1.1)(100) + (1.1)^2(100) + \ldots$
$\qquad + (1.1)^{18}(100) + (1.1)^{19}(100)$ pounds

Using the formula for the sum of a G.P.

$$\frac{a(r^n - 1)}{r - 1}$$

where $a \equiv$ first term; $n \equiv$ number of terms and $r \equiv$ common ratio with $a = 100$, $n = 20$ and $r = 1.1$ we have

$$\frac{100(1.1^{20} - 1)}{1.1 - 1} = \frac{100(1.1^{20} - 1)}{0.1} = 5727.5$$

Therefore the total investment is £5730 to the nearest £10.

We can represent a complex number of the form $z = x + iy$ on an Argand Diagram as follows:

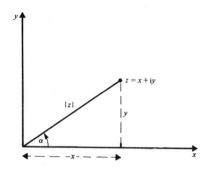

The modulus of z, $|z|$, is represented by the distance from the origin to the point z. The argument of z, arg (z), is the angle α, where $-\pi \leqslant \alpha \leqslant \pi$.

$z_1 = 1 - i$

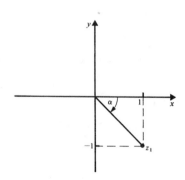

$$|z_1| = \sqrt{1^2 + 1^2} = \sqrt{2}$$

$$\arg(z_1) = -\alpha, \quad \text{where} \quad \tan \alpha = \frac{1}{1} = 1$$

$\Rightarrow \qquad \alpha = \dfrac{\pi}{4}$

$\Rightarrow \qquad \arg(z_1) = -\dfrac{\pi}{4}$

$$\begin{aligned}
z_2 = (1-i)^3 &= (1-i)^2(1-i) \\
&= (1-i-i+i^2)(1-i) \\
&= (1-2i-1)(1-i) \\
&= (-2i)(1-i) \\
&= -2i + 2i^2
\end{aligned}$$

$\Rightarrow \qquad z_2 = -2i - 2$

34

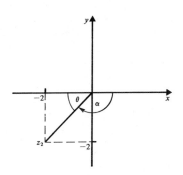

$$|z_2| = \sqrt{2^2 + 2^2} = \sqrt{8} = \sqrt{4 \times 2} = 2\sqrt{2}$$

$$\arg(z_2) = -\alpha, \quad \text{where} \quad \alpha = \pi - \theta$$

$$\text{and} \quad \tan\theta = \frac{2}{2} = 1$$

$$\Rightarrow \qquad \theta = \frac{\pi}{4}$$

$$\Rightarrow \qquad \arg(z_2) = -\frac{3}{4}\pi$$

To put z_3 into the form $x + iy$ we multiply the numerator and the denominator by the complex conjugate of the denominator. This eliminates i from the denominator.

$$z_3 = \frac{\sqrt{3} - i}{\sqrt{3} + i} = \frac{(\sqrt{3} - i)(\sqrt{3} - i)}{(\sqrt{3} + i)(\sqrt{3} - i)} = \frac{3 - \sqrt{3}i - \sqrt{3}i + i^2}{3 - \sqrt{3}i + \sqrt{3}i - i^2}$$

$$= \frac{3 - 2\sqrt{3}i - 1}{3 - i^2} = \frac{2 - 2\sqrt{3}i}{3 + 1} = \frac{2 - 2\sqrt{3}i}{4}$$

$$\Rightarrow \qquad z_3 = \frac{1}{2} - \frac{\sqrt{3}}{2}i$$

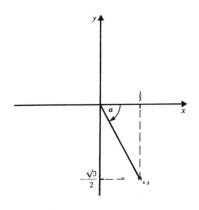

$$|z_3| = \sqrt{\left(\frac{1}{2}\right)^2 + \left(\frac{\sqrt{3}}{2}\right)^2} = \sqrt{\frac{1}{4} + \frac{3}{4}} = \sqrt{1} = 1$$

$$\arg(z_3) = -\alpha, \quad \text{where} \quad \tan\alpha = \frac{\sqrt{3}/2}{\frac{1}{2}} = \frac{\sqrt{3}}{1} = \sqrt{3}$$

$$\Rightarrow \qquad \alpha = \frac{\pi}{3}$$

$$\Rightarrow \qquad \arg(z_3) = -\frac{\pi}{3}$$

Marking all three points on the same diagram gives

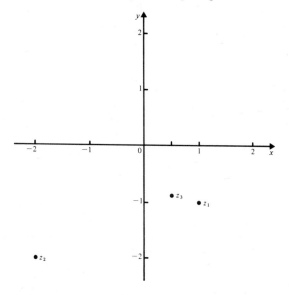

10

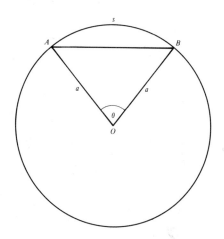

(a) To find the area of triangle $OAB \equiv \Delta_1$ we use the formula

$$\text{Area} = \tfrac{1}{2}(OA),(OB),(\sin A\widehat{O}B) \qquad [1]$$

Therefore we must find angle $A\widehat{O}B$.

36

Arc length AB is $\dfrac{\theta}{2\pi}$ of the circumference $(2\pi a)$

$$\Rightarrow \quad s = \frac{\theta}{2\pi} \times 2\pi a = \theta a$$

$$\Rightarrow \quad \theta = \frac{s}{a} \qquad\qquad [2]$$

Therefore from [1]

$$\Delta_1 = \tfrac{1}{2}a^2 \sin\left(\frac{s}{a}\right) \qquad\qquad [3]$$

(b) Sector $OAB \equiv \Delta_2$ is $\dfrac{\theta}{2\pi}$ of the total area of the circle (πa^2)

$$\Rightarrow \quad \Delta_2 = \frac{\theta}{2\pi} \times \pi a^2 = \tfrac{1}{2}\theta a^2$$

$$= \frac{1}{2}\left(\frac{s}{a}\right)a^2 \quad \text{from [2]}$$

$$\Rightarrow \quad \Delta_2 = \tfrac{1}{2}sa \qquad\qquad [4]$$

With simple trigonometric functions we do not need to differentiate to find maximum values since sin x and cos x both have maximum values of 1.

$$\Delta_1 = \tfrac{1}{2}a^2 \sin\left(\frac{s}{a}\right) \quad \text{equation [3]}$$

$\Rightarrow \quad \Delta_1$ has a maximum value when $\sin\left(\dfrac{s}{a}\right)$ is a maximum

$$\Rightarrow \quad \sin\left(\frac{s}{a}\right) = 1$$

$\Rightarrow \quad \Delta_1$ is maximum when $\dfrac{s}{a} = \arcsin 1$

$$\Rightarrow \quad \frac{s}{a} = \frac{\pi}{2}$$

It is not necessary to consider other values of $\dfrac{s}{a}$ because θ

$\left(\text{which equals } \dfrac{s}{a}\right)$ *must be between 0 and 2π.*

$\Rightarrow \quad \Delta_1$ is a maximum when $\dfrac{s}{a} = \dfrac{\pi}{2}$

$$\Rightarrow \quad s = \frac{a\pi}{2} \qquad\qquad [5]$$

When $s = \dfrac{a\pi}{2}$

(i) $\Delta_1 = \frac{1}{2}a^2 \sin\left(\dfrac{a\pi/2}{a}\right)$ from [3]

$\Rightarrow$ $\Delta_1 = \frac{1}{2}a^2 \sin\left(\dfrac{\pi}{2}\right)$

$\Rightarrow$ $\Delta_1 = \frac{1}{2}a^2 \left(\text{since} \quad \sin\dfrac{\pi}{2} = 1\right)$ [6]

and

(ii) $\Delta_2 = \frac{1}{2}sa$ equation [4]

$\Rightarrow$ $\Delta_2 = \dfrac{1}{2}\left(\dfrac{a\pi}{2}\right)a$ from [5]

$\Rightarrow$ $\Delta_2 = \frac{1}{4}\pi a^2$ [7]

Therefore from [6] and [7]

$$\pi\Delta_1 = 2\Delta_2 \quad \text{when} \quad s = \dfrac{a\pi}{2}$$

11

When sketching curves of this type we should consider the following:

  (i) does the curve cross either axis (i.e. what happens when $x = 0$ and $y = 0$)?
 (ii) are there any values of x for which the denominator is zero (i.e. are there any vertical asymptotes)?
(iii) the values of y for large values of x — both positive and negative (i.e. are there any horizontal asymptotes?).
 (iv) how does the curve approach its asymptotes?

In this type of question the emphasis is more on investigation than algebraic technique. It is very important that we explain clearly how the function behaves at each of the steps (i) to (iv), above. We can then mark this information on a diagram (e.g. Figures 1 and 2, opposite and overleaf) to produce a skeleton sketch. It will then become clear how the sketch may be completed (see Figure 3, overleaf).

For $y = \dfrac{1}{x - 1}$

 (i) (a) When $x = 0$, $y = -1$.
     (b) There is no value of x for which $y = 0$.

 (ii) $x - 1 = 0$ when $x = 1$
      $\Rightarrow$ $x = 1$ is a vertical asymptote.

(iii) (a) as $x \to +\infty$, $y \to 0$ (and is positive)
      (b) as $x \to -\infty$, $y \to 0$ (and is negative)
          $\Rightarrow$ $y = 0$ is a horizontal asymptote.

(iv) We now check values of x close to either side of the vertical asymptote, say $x = 1.1$ and $x = 0.9$.

When $x = 1.1$, $y = \dfrac{1}{1.1 - 1} = \dfrac{1}{0.1} = 10$

(i.e. large and positive)

When $x = 0.9$, $y = \dfrac{1}{0.9 - 1} = \dfrac{1}{-0.1} = -10$

(i.e. large and negative)

This information produces Figure 1.

$x = 1$ **Figure 1**

For $y = \dfrac{x}{x + 3}$

(i) (a) When $x = 0$, $y = 0$,   (b) when $y = 0$, $x = 0$.

(ii) $x + 3 = 0$ when $x = -3$
   $\Rightarrow$ $x = -3$ is a vertical asymptote.

(iii) (a) As $x \to +\infty$, $y \to 1$ and because $x + 3 > x$, y will be just under 1.
   (b) As $x \to -\infty$, $y \to 1$ and because $|x + 3| < |x|$, y will be just over 1
     $\Rightarrow$ $y = 1$ is a horizontal asymptote.

(iv) We now check values of x close to either side of the vertical asymptote, say $x = -3.1$ and $x = -2.9$.

When $x = -3.1$, $y = \dfrac{-3.1}{-3.1 + 3} = \dfrac{-3.1}{-0.1} = 31$

(i.e. large and positive)

When $x = -2.9$, $y = \dfrac{-2.9}{-2.9 + 3} = \dfrac{-2.9}{0.1} = -29$

(i.e. large and negative)

This information produces Figure 2.

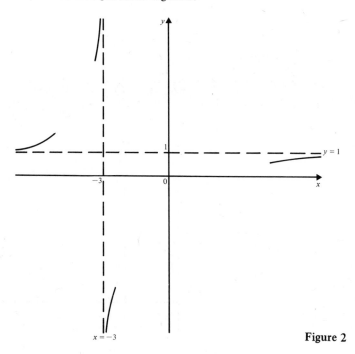

Figure 2

If we combine Figures 1 and 2 we produce Figure 3.

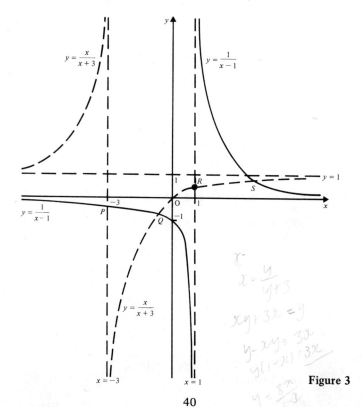

Figure 3

40

We can find the set of values of x for which $\dfrac{1}{x-1} > \dfrac{x}{x+3}$

by considering the places in Figure 3, opposite where the graph of $y = \dfrac{1}{x-1}$ is above the graph of $y = \dfrac{x}{x+3}$, i.e. for values of x between points P and Q, and R and S.

As we can see from the diagram the x-value of P is -3 and the x-value of R is 1.

Q and S are the points of intersection of the two functions, i.e. when

$$\frac{x}{x+3} = \frac{1}{x-1}$$

Solving this equation gives

$$x(x-1) = (x+3)$$
$$\Rightarrow \qquad x^2 - x = x + 3$$
$$\Rightarrow \qquad x^2 - 2x - 3 = 0$$
$$\Rightarrow \qquad (x-3)(x+1) = 0$$

either $x - 3 = 0$ or $x + 1 = 0$
$$\Rightarrow \qquad\qquad x = 3 \quad \text{or} \quad -1$$
$\Rightarrow \qquad$ The x-values of Q and S are -1 and 3 respectively.

Hence $\dfrac{1}{x-1} > \dfrac{x}{x+3}$ for

$$\{x: -3 < x < -1\} \cup \{x: 1 < x < 3\}$$

12

As any curve, $f(x)$, crosses the x-axis the value of $f(x)$ will change from positive to negative, or vice versa.

$$f(-2) = -8 + 4 + 4 - 1 = -1$$
$$f(-1) = -1 + 1 + 2 - 1 = 1$$
$$f(0) = 0 + 0 - 0 - 1 = -1$$
$$f(1) = 1 + 1 - 2 - 1 = -1$$
$$f(2) = 8 + 4 - 4 - 1 = 7$$

From the above information we see that $f(x)$ changes sign between

(a) $x = -2$ and $x = -1$,

(b) $x = -1$ and $x = 0$, and

(c) $x = 1$ and $x = 2$.

Therefore there must be a value of x in each of these intervals where $f(x) = 0$.

Therefore $f(x) = 0$ has a root in each of the intervals

$$x < -1, \quad -1 < x < 0 \quad \text{and} \quad x > 1$$

$\Rightarrow \qquad$ The only positive root is in the interval $x > 1$.

In order to use the Newton–Raphson procedure we need $f'(x)$

$$f(x) = x^3 + x^2 - 2x - 1$$
$$\Rightarrow \quad f'(x) = 3x^2 + 2x - 2$$

The Newton–Raphson procedure states that if $x = 1$ is an approximate solution to $f(x) = 0$ then

$$x_1 = 1 - \frac{f(1)}{f'(1)} \qquad [1]$$

is a better approximation.

$$f(1) = 1 + 1 - 2 - 1 = -1$$

and

$$f'(1) = 3 + 2 - 2 = 3$$
$$\Rightarrow \quad x_1 = 1 - \quad -\tfrac{1}{3} \quad \text{from } [1]$$
$$\Rightarrow \quad x_1 = 1\tfrac{1}{3} \quad \text{is a better approximation}$$

A second application of the procedure gives

$$x_2 = 1\tfrac{1}{3} - \frac{f(1\tfrac{1}{3})}{f'(1\tfrac{1}{3})} \qquad [2]$$

as a better approximation than x_1.

$$f(1\tfrac{1}{3}) = \tfrac{64}{27} + \tfrac{16}{9} - \tfrac{8}{3} - 1 = \tfrac{13}{27}$$

and

$$f'(1\tfrac{1}{3}) = \tfrac{16}{3} + \tfrac{8}{3} - 2 = 6$$
$$\Rightarrow \quad x_2 = 1\tfrac{1}{3} - \frac{\tfrac{13}{27}}{6} \quad \text{from } [2]$$
$$= 1\tfrac{1}{3} - \tfrac{13}{162} = 1\tfrac{41}{162}$$
$$\Rightarrow \quad x_2 = 1.25 \text{ to 2 decimal places}$$

Therefore the positive root of $f(x) = 0$ is $x = 1.25$.

13

Let $$\frac{2}{(1 + x)(1 + 3x)} \equiv \frac{A}{(1 + x)} + \frac{B}{(1 + 3x)} \qquad [1]$$

Therefore

$$\frac{2}{(1 + x)(1 + 3x)} \equiv \frac{A(1 + 3x) + B(1 + x)}{(1 + x)(1 + 3x)}$$

If we compare the numerators, then

$$2 \equiv A(1 + 3x) + B(1 + x) \qquad [2]$$

To find A we will make B 'disappear' by putting $x = -1$ in [2]

$$\Rightarrow \quad 2 = A(1 + -3) + B(1 + -1)$$
$$\Rightarrow \quad 2 = -2A$$
$$\Rightarrow \quad A = -1$$

To find B we will make A 'disappear' by putting $x = -\frac{1}{3}$ in [2]

$$2 = A(1 + -1) + B(1 + -\tfrac{1}{3})$$

$\Rightarrow \quad 2 = \tfrac{2}{3}B$

$\Rightarrow \quad B = 3$

Hence, substituting $A = -1$ and $B = 3$ into [1] gives

$$\frac{2}{(1+x)(1+3x)} \equiv \frac{-1}{(1+x)} + \frac{3}{(1+3x)} \qquad [3]$$

To solve the differential equation we will separate dx from dy and collect terms in y and dy on the L.H.S. and terms in x and dx on the R.H.S. Then we shall integrate each side.

$$\frac{dy}{dx} = \frac{2(y+2)}{(1+x)(1+3x)}$$

$\Rightarrow \qquad dy = \dfrac{2(y+2)\,dx}{(1+x)(1+3x)}$

$\Rightarrow \qquad \dfrac{dy}{(y+2)} = \dfrac{2\,dx}{(1+x)(1+3x)}$

$\Rightarrow \quad \displaystyle\int \frac{dy}{(y+2)} = \int \frac{2\,dx}{(1+x)(1+3x)}$

We now express the R.H.S. as the partial fractions found in [3].

$\Rightarrow \quad \displaystyle\int \frac{dy}{(y+2)} = \int \left(\frac{-1}{(1+x)} + \frac{3}{(1+3x)} \right) dx$

$\Rightarrow \quad \ln(y+2) = -\ln(1+x) + \ln(1+3x) + \ln A$

In integration involving logarithms the constant of integration may also be taken as a logarithm.

$\Rightarrow \quad \ln(y+2) = \ln\left(\dfrac{1+3x}{1+x}\right) + \ln A = \ln A\left(\dfrac{1+3x}{1+x}\right)$

by the laws of logarithms

$\Rightarrow \qquad y + 2 = A\left(\dfrac{1+3x}{1+x}\right) \qquad [4]$

To find the value of A we substitute the given information
$y = -1$ when $x = 0$

$\Rightarrow \qquad -1 + 2 = A\left(\dfrac{1+0}{1+0}\right)$

$\Rightarrow \qquad 1 = A$

Substituting $A = 1$ into [4] gives

$$y + 2 = \frac{1+3x}{1+x}$$

$$\Rightarrow \qquad y = \frac{(1 + 3x)}{(1 + x)} - 2$$

$$= \frac{1 + 3x - 2(1 + x)}{(1 + x)}$$

$$= \frac{1 + 3x - 2 - 2x}{(1 + x)}$$

$$= \frac{-1 + x}{(1 + x)}$$

Therefore $y = \dfrac{x - 1}{x + 1}$ is the solution to the differential equation.

14

To find the greatest and least values of this trigonometric function we will use differentiation.

Let $\quad y = \dfrac{\sin x}{2 - \cos x}$

The Quotient Rule for differentiation is

$$\frac{dy}{dx} = \frac{v\dfrac{du}{dx} - u\dfrac{dv}{dx}}{v^2} \qquad [1]$$

In this case

$$u = \sin x \quad \text{and} \quad v = 2 - \cos x$$

$$\Rightarrow \quad \frac{du}{dx} = \cos x \quad \text{and} \quad \frac{dv}{dx} = \sin x$$

Substituting into [1] gives

$$\frac{dy}{dx} = \frac{(2 - \cos x)(\cos x) - (\sin x)(\sin x)}{(2 - \cos x)^2}$$

$$\Rightarrow \quad \frac{dy}{dx} = \frac{2 \cos x - \cos^2 x - \sin^2 x}{(2 - \cos x)^2}$$

$$= \frac{2 \cos x - (\cos^2 x + \sin^2 x)}{(2 - \cos x)^2}$$

$$\Rightarrow \quad \frac{dy}{dx} = \frac{2 \cos x - 1}{(2 - \cos x)^2} \quad (\text{since } \cos^2 x + \sin^2 x = 1) \qquad [2]$$

The maximum and minimum values of $y \equiv f(x)$ occur when $\dfrac{dy}{dx} = 0$.

$\dfrac{dy}{dx} = 0$ when the numerator only is 0.

$\Rightarrow \qquad 2\cos x - 1 = 0 \qquad$ from [2]

$\Rightarrow \qquad\qquad 2\cos x = 1$

$\Rightarrow \qquad\qquad\quad \cos x = \tfrac{1}{2}$

$\Rightarrow \qquad\qquad\qquad x = \dfrac{\pi}{3} \quad \text{and} \quad \dfrac{5\pi}{3}$

When $x = \dfrac{\pi}{3}$

$$f(x) = \frac{\sin \frac{\pi}{3}}{2 - \cos \frac{\pi}{3}} = \frac{\frac{\sqrt{3}}{2}}{2 - \frac{1}{2}} = \frac{\frac{\sqrt{3}}{2}}{\frac{3}{2}} = \frac{\sqrt{3}}{3}$$

When $x = \dfrac{5\pi}{3}$

$$f(x) = \frac{\sin \frac{5\pi}{3}}{2 - \cos \frac{5\pi}{3}} = \frac{-\frac{\sqrt{3}}{2}}{2 - \frac{1}{2}} = -\frac{\frac{\sqrt{3}}{2}}{\frac{3}{2}} = -\frac{\sqrt{3}}{3}$$

Therefore the greatest value of $f(x)$ must be $\dfrac{\sqrt{3}}{3}$

and the least value of $f(x)$ must be $-\dfrac{\sqrt{3}}{3}$.

(N.B. *It is not necessary to differentiate twice to distinguish between the maximum and minimum values.*)

Before we integrate f(x) to find the area we must ensure that the curve y = f(x) does not cross the x-axis between the limits x = 0 and x = π.

We know $f(x)$ has only one stationary point between $x = 0$ and $x = \pi$ $\left(\text{a maximum at } x = \dfrac{\pi}{3}\right)$ and that $f(x) = 0$ when $x = 0$ and $x = \pi$. We also know that the curve is continuous because the denominator of $f(x)$ can never be zero, since $\cos x$ can never equal 2.

Hence the curve is always positive between $x = 0$ and $x = \pi$.

Therefore it does not cross the x-axis.

The required area, A, is given by

$$A = \int_0^{\pi} f(x)\, dx$$

$$\Rightarrow \quad A = \int_0^{\pi} \frac{\sin x}{2 - \cos x}\, dx$$

As the denominator differentiates to give the numerator

$$A = \Big[\ln(2 - \cos x) \Big]_0^{\pi}$$

45

$\Rightarrow \quad A = \ln(2 - \cos \pi) - \ln(2 - \cos 0)$

$\qquad = \ln(2 - -1) - \ln(2 - 1)$

$\qquad = \ln 3 - \ln 1$

$\Rightarrow \quad A = \ln 3 \quad$ (since $\quad \ln 1 = 0$)

Therefore the area bounded by $y = f(x)$ and the x-axis between $x = 0$ and $x = \pi$ is $\ln 3$.

To show that the unit vector j *is perpendicular to the plane of the triangle LMN we need to show that* j *is perpendicular to two sides of the triangle. We can do this by showing that the scalar products* $j . \overrightarrow{LM}$ *and* $j . \overrightarrow{LN}$ *are both zero.*

$\qquad \overrightarrow{LM} = \mathbf{m} - \mathbf{l}$

$\qquad\qquad = a(2\mathbf{i} + \mathbf{j}) - a(\mathbf{i} + \mathbf{j} + \mathbf{k})$

$\qquad\qquad = 2a\mathbf{i} + a\mathbf{j} - a\mathbf{i} - a\mathbf{j} - a\mathbf{k}$

$\qquad\qquad = a\mathbf{i} - a\mathbf{k}$

$\Rightarrow \quad \overrightarrow{LM} = a(\mathbf{i} - \mathbf{k})$

$\qquad \overrightarrow{LN} = \mathbf{n} - \mathbf{l}$

$\qquad\qquad = a(\mathbf{j} + 4\mathbf{k}) - a(\mathbf{i} + \mathbf{j} + \mathbf{k})$

$\qquad\qquad = a\mathbf{j} + 4a\mathbf{k} - a\mathbf{i} - a\mathbf{j} - a\mathbf{k}$

$\qquad\qquad = -a\mathbf{i} + 3a\mathbf{k}$

$\Rightarrow \quad \overrightarrow{LN} = a(-\mathbf{i} + 3\mathbf{k})$

Therefore $\qquad \mathbf{j} . \overrightarrow{LM} = \mathbf{j} . a(\mathbf{i} - \mathbf{k}) = 0$

and $\qquad\qquad \mathbf{j} . \overrightarrow{LN} = \mathbf{j} . a(-\mathbf{i} + 3\mathbf{k}) = 0$

since

$\qquad \mathbf{j} . \mathbf{i} \quad$ and $\quad \mathbf{j} . \mathbf{k} \quad$ both equal zero

Therefore $\mathbf{j}$ is perpendicular to triangle *LMN*.

Let $x\mathbf{i} + y\mathbf{j} + z\mathbf{k}$ be a vector perpendicular to both $\mathbf{j}$ and $(\mathbf{m} - \mathbf{n})$.

Therefore

$\qquad\qquad \mathbf{j} . (x\mathbf{i} + y\mathbf{j} + z\mathbf{k}) = 0 \qquad\qquad$ [1]

and $\quad (\mathbf{m} - \mathbf{n}) . (x\mathbf{i} + y\mathbf{j} + z\mathbf{k}) = 0 \qquad\qquad$ [2]

From [1]

$\qquad\qquad \mathbf{j} . (x\mathbf{i} + y\mathbf{j} + z\mathbf{k}) = 0$

$\Rightarrow \qquad\qquad\qquad\qquad y = 0 \qquad\qquad$ [3]

To evaluate [2] we must first evaluate $(\mathbf{m} - \mathbf{n})$.

$\qquad (\mathbf{m} - \mathbf{n}) = a(2\mathbf{i} + \mathbf{j}) - a(\mathbf{j} + 4\mathbf{k})$

$\qquad\qquad\qquad = 2a\mathbf{i} + a\mathbf{j} - a\mathbf{j} - 4a\mathbf{k}$

$\qquad\qquad\qquad = 2a\mathbf{i} - 4a\mathbf{k}$

$\Rightarrow \quad (m - n) = a(2\mathbf{i} - 4\mathbf{k})$

Substituting into [2] gives

$$a(2\mathbf{i} - 4\mathbf{k}).(x\mathbf{i} + y\mathbf{j} + z\mathbf{k}) = 0$$

$\Rightarrow \qquad\qquad\qquad\qquad 2ax - 4az = 0$

$\Rightarrow \qquad\qquad\qquad\qquad\qquad 2ax = 4az$

$\Rightarrow \qquad\qquad\qquad\qquad\qquad x = 2z \quad \text{(since } a \text{ is non-zero)} \quad [4]$

Therefore

$$(x\mathbf{i} + y\mathbf{j} + z\mathbf{k}) \equiv (2z\mathbf{i} + 0\mathbf{j} + z\mathbf{k})$$

$$\text{from [3] and [4]}$$

$$\equiv z(2\mathbf{i} + \mathbf{k}).$$

Therefore a vector perpendicular to both $\mathbf{j}$ and $(\mathbf{m} - \mathbf{n})$ is

$$2\mathbf{i} + \mathbf{k}$$

The vector $2\mathbf{i} + \mathbf{k}$ must lie in the plane of the triangle because

(i) it is perpendicular to $(\mathbf{m} - \mathbf{n})$ which represents the side $\overrightarrow{NM}$ of the triangle, and

(ii) it is also perpendicular to $\mathbf{j}$, a vector perpendicular to the plane of the triangle.

Therefore a vector equation of the perpendicular bisector of MN will be $\mathbf{r} = \mathbf{p} + t(2\mathbf{i} + \mathbf{k})$ where $\mathbf{p}$ is the position vector of the mid point of MN.

$\Rightarrow \qquad \mathbf{r} = \frac{1}{2}\mathbf{m} + \frac{1}{2}\mathbf{n} + t(2\mathbf{i} + \mathbf{k})$

(since $\mathbf{p} = \frac{1}{2}\mathbf{m} + \frac{1}{2}\mathbf{n}$ by the Ratio Theorem)

$\Rightarrow \qquad \mathbf{r} = \frac{1}{2}a(2\mathbf{i} + \mathbf{j}) + \frac{1}{2}a(\mathbf{j} + 4\mathbf{k}) + t(2\mathbf{i} + \mathbf{k})$

$\qquad\qquad = \frac{1}{2}a(2\mathbf{i} + \mathbf{j} + \mathbf{j} + 4\mathbf{k}) + t(2\mathbf{i} + \mathbf{k})$

$\qquad\qquad = \frac{1}{2}a(2\mathbf{i} + 2\mathbf{j} + 4\mathbf{k}) + t(2\mathbf{i} + \mathbf{k})$

$\Rightarrow \qquad \mathbf{r} = a(\mathbf{i} + \mathbf{j} + 2\mathbf{k}) + t(2\mathbf{i} + \mathbf{k})$

We can further change $\mathbf{r}$ into

$$\mathbf{r} = a\mathbf{i} + a\mathbf{j} + 2a\mathbf{k} + 2t\mathbf{i} + t\mathbf{k}$$

$\Rightarrow \qquad \mathbf{r} = (a + 2t)\mathbf{i} + a\mathbf{j} + (2a + t)\mathbf{k}$

For the point K with position vector $(5a\mathbf{i} + a\mathbf{j} + 4a\mathbf{k})$ to lie on $\mathbf{r}$ we must find the value of t so that

$$a + 2t = 5a$$

$\Rightarrow \qquad\qquad 2t = 4a$

$\Rightarrow \qquad\qquad\; t = 2a$

and $\quad 2a + t = 4a$

$\Rightarrow \qquad\qquad\; t = 2a$

$\Rightarrow \qquad K$ lies on this bisector

Let the position vector of K be $\mathbf{q}$ (because $\mathbf{k}$ is used for a unit vector)

$$\overrightarrow{KL} = 1 - \mathbf{q} = a(\mathbf{i} + \mathbf{j} + \mathbf{k}) - a(5\mathbf{i} + \mathbf{j} + 4\mathbf{k})$$

$$= a\mathbf{i} + a\mathbf{j} + a\mathbf{k} - 5a\mathbf{i} - a\mathbf{j} - 4a\mathbf{k}$$

$$= -4a\mathbf{i} - 3a\mathbf{k}$$

$$\Rightarrow \quad |\overrightarrow{KL}| = \sqrt{(-4a)^2 + (-3a)^2} = \sqrt{16a^2 + 9a^2} = \sqrt{25a^2}$$
$$= 5a$$

$$\overrightarrow{KM} = \mathbf{m} - \mathbf{q} = a(2\mathbf{i} + \mathbf{j}) - a(5\mathbf{i} + \mathbf{j} + 4\mathbf{k})$$
$$= 2a\mathbf{i} + a\mathbf{j} - 5a\mathbf{i} - a\mathbf{j} - 4a\mathbf{k}$$
$$= -3a\mathbf{i} - 4a\mathbf{k}$$

$$\Rightarrow \quad |\overrightarrow{KM}| = \sqrt{(-3a)^2 + (-4a)^2} = \sqrt{9a^2 + 16a^2} = \sqrt{25a^2}$$
$$= 5a$$

$$\overrightarrow{KN} = \mathbf{n} - \mathbf{q} = a(\mathbf{j} + 4\mathbf{k}) - a(5\mathbf{i} + \mathbf{j} + 4\mathbf{k})$$
$$= a\mathbf{j} + 4a\mathbf{k} - 5a\mathbf{i} - a\mathbf{j} - 4a\mathbf{k}$$
$$= -5a\mathbf{i}$$

$$\Rightarrow \quad |\overrightarrow{KN}| = 5a$$

Therefore k is equidistant from L, M and N.

January 1984

PAPER 2

1

We consider the three post-boxes P, Q and R as being 'fixed'.

The number of ways of posting 4 of the 8 letters into box P is $^8C_4$.

There are only 4 letters remaining.

The number of ways of posting 2 of these 4 letters into box Q is $^4C_2$.

The remaining letters *must* go into box R.

This can only be done in 1 way.

Therefore the total number of ways of posting all 8 letters is

$$^8C_4 \times {}^4C_2 \times 1 = \frac{8!}{4! \times 4!} \times \frac{4!}{2! \times 2!} \times 1$$

$$= 70 \times 6 \times 1$$

$$= 420$$

2

By definition the arithmetic mean of p and q is $\dfrac{p+q}{2}$ *and the geometric mean is* $\sqrt{pq}$.

Hence

$$\frac{p+q}{2} = 39$$

$$\Rightarrow \quad p + q = 78 \tag{1}$$

and $\quad \sqrt{pq} = 15$

$$\Rightarrow \quad pq = 225 \tag{2}$$

If p and q are the roots of a quadratic equation then in [1] and [2] we have found the sum and product of these roots.

The general form of a quadratic equation is

$$ax^2 + bx + c = 0$$

If this equation has roots α and β and we write it as:

$$x^2 + \frac{b}{a}x + \frac{c}{a} = 0 \tag{3}$$

then the sum of the roots

$$\alpha + \beta = -\frac{b}{a} \tag{4}$$

and the product of the roots

$$\alpha\beta = \frac{c}{a} \tag{5}$$

$$p + q = 78 \qquad \text{equation [1]}$$

$$\Rightarrow \quad -\frac{b}{a} = 78 \qquad \text{from [4]}$$

$$\text{and} \quad pq = 225 \qquad \text{equation [2]}$$

$$\Rightarrow \quad \frac{c}{a} = 225 \qquad \text{from [5]}$$

Therefore, combining this information with [3], the required quadratic equation is:

$$x^2 - 78x + 225 = 0 \qquad [6]$$

To find p and q we can either solve [1] and [2] simultaneously or solve [6] as a quadratic equation. Since the first method will lead to another quadratic equation we will find p and q using [6].

$$x^2 - 78x + 225 = 0$$

$$\Rightarrow \quad (x - 75)(x - 3) = 0$$

$$\Rightarrow \quad \text{either} \quad x - 75 \quad \text{or} \quad x - 3 = 0$$

$$\Rightarrow \quad x = 75 \quad \text{and} \quad x = 3 \quad \text{are the two roots}$$

Therefore

$$p = 75 \quad \text{and} \quad q = 3 \quad \text{or} \quad q = 75 \quad \text{and} \quad p = 3.$$

3

To evaluate this integral we use the formula for Integration by Parts:

$$\int v \frac{du}{dx} dx = uv - \int u \frac{dv}{dx} dx \qquad [1]$$

with

$$v = x \quad \text{and} \quad \frac{du}{dx} = \sin x$$

$$\Rightarrow \quad \frac{dv}{dx} = 1 \quad \text{and} \quad u = -\cos x$$

Substituting into [1] gives

$$\int_0^{2\pi} x(\sin x)\, dx = \left[(-\cos x)(x)\right]_0^{2\pi} - \int_0^{2\pi} (-\cos x)(1)\, dx$$

$$\Rightarrow \quad \int_0^{2\pi} x(\sin x)\, dx = \left[-x(\cos x)\right]_0^{2\pi} + \int_0^{2\pi} \cos x\, dx$$

$$= \left[-x(\cos x)\right]_0^{2\pi} + \left[\sin x\right]_0^{2\pi}$$

$$= (-2\pi \cos 2\pi - -0 \cos 0)$$

$$+ (\sin 2\pi - \sin 0)$$

$$= -2\pi + 0$$

Therefore

$$\int_0^{2\pi} x(\sin x)\, dx = -2\pi$$

$$|x - 2| - 2|2x - 1| > 0$$

$$\Rightarrow \quad |x - 2| > 2|2x - 1|$$

Let $f(x) = |x - 2|$ and $g(x) = 2|2x - 1|$

We will sketch the two functions f(x) and g(x) on the same pair of axes and where f(x) lies above g(x) will be the solution to the inequality:

$$|x - 2| > 2|2x - 1|$$

$f(x) = 0$ when $x = 2$ and by definition $f(x) \geq 0$ and is symmetrical about the line $x = 2$.

We now find the value of $f(x)$ at two points either side of $x = 2$ and then sketch the function $f(x)$.

When $x = 1$

$$f(x) = |1 - 2| = |-1| = 1$$

When $x = 3$

$$f(x) = |3 - 2| = |1| = 1$$

We are now able to sketch $f(x)$.

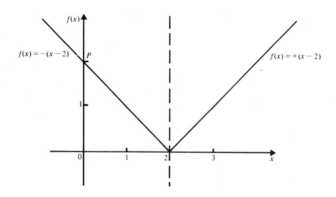

We can find the coordinates of the point P by evaluating $f(0)$.

$$f(0) = |0 - 2| = |-2| = 2$$

Therefore P is the point $(0, 2)$.

$g(x) = 0$ when $x = \frac{1}{2}$ and by definition $g(x) \geq 0$ and is symmetrical about the line $x = \frac{1}{2}$.

We now find the value of $g(x)$ at two points either side of $x = \frac{1}{2}$ and then sketch the function $g(x)$.

When $x = 1$

$$g(x) = 2|2 - 1| = 2|1| = 2$$

When $x = 0$

$$g(x) = 2|0 - 1| = 2|-1| = 2$$

We are now able to sketch $g(x)$.

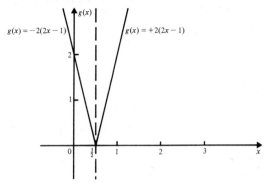

We can now sketch the two functions on the same diagram.

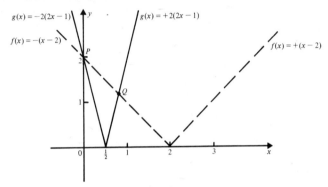

From the diagram, $f(x) > g(x)$ for x between the two intersection points P and Q. At P, $x = 0$ and at Q the *negative branch* of $f(x)$ meets the *positive branch* of $g(x)$.

Therefore

$$-(x - 2) = 2(2x - 1)$$
$$\Rightarrow \quad -x + 2 = 4x - 2$$
$$\Rightarrow \quad 2 = 5x - 2$$
$$\Rightarrow \quad 4 = 5x$$
$$\Rightarrow \quad x = \tfrac{4}{5}$$

Therefore

$$f(x) > g(x) \text{ and } |x - 2| - 2|2x - 1| > 0 \text{ for } \{x: 0 < x < \tfrac{4}{5}\}$$

5

When $x = 0$

$$f(x) - 4(0)^2 = 0$$

When $x = \dfrac{\pi}{4}$

$$f(x) = 4\left(\frac{\pi}{4}\right)^2 = \frac{4\pi^2}{16} = \frac{\pi^2}{4}$$

53

Therefore for x in the range $0 \leqslant x \leqslant \dfrac{\pi}{4}$ the sketch is Figure 1.

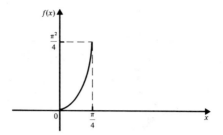

Figure 1

For x in the range $\dfrac{\pi}{4} < x \leqslant \dfrac{\pi}{2}$, $f(x) = \dfrac{\pi^2}{4}$ and is therefore constant. Figure 1 can now be extended to:

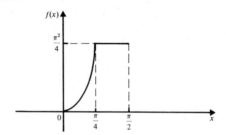

Figure 2

An even function has the y-axis as a line of symmetry and so Figure 2 can be reflected to give Figure 3.

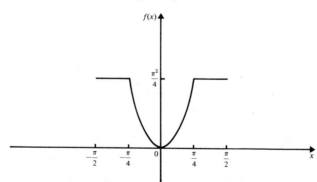

Figure 3

A periodic function, with period π, will repeat the same shape every π units along the x-axis.

$$\frac{-\pi}{2} \leqslant x \leqslant \frac{\pi}{2} \quad \text{is a complete period of } \pi$$

Therefore Figure 3 can be repeated in both negative and positive directions (but only as far as $x = -\pi$ and $x = \pi$ for this question).

Therefore the complete sketch of $f(x)$ in the range $-\pi \leqslant x \leqslant \pi$ is Figure 4.

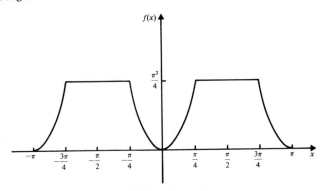

Figure 4

6

The sum of $(3i + 2j + k)$ and $(-5i - 3j + 6k)$ is

$$(3 - 5)i + (2 - 3)j + (1 + 6)k = -2i - j + 7k$$

The vector in the opposite direction is

$$-(-2i - j + 7k) = 2i + j - 7k$$

If $v = ai + bj + ck$ is a vector then the unit vector in the direction v is

$$\frac{1}{|v|}(ai + bj + ck), \quad \text{where} \quad |v| = \sqrt{\{a^2 + b^2 + c^2\}}$$

With $a = 2$, $b = 1$ and $c = -7$ the unit vector in the direction $2i + j - 7k$ is

$$\frac{1}{|v|}(2i + j - 7k) \quad \text{where} \quad |v| = \sqrt{\{2^2 + 1^2 + (-7)^2\}}$$
$$= \sqrt{\{4 + 1 + 49\}} = \sqrt{54}$$

Hence the unit vector is

$$\frac{1}{\sqrt{54}}(2i + j - 7k)$$

If two vectors p and q are perpendicular then the scalar product of p and q is zero.

$$\frac{1}{\sqrt{54}}(2i + j - 7k) \cdot (9i - 4j + 2k)$$

$$= \frac{1}{\sqrt{54}}[2 \times 9 + 1 \times (-4) + (-7) \times 2]$$

$$= \frac{1}{\sqrt{54}}(18 - 4 - 14)$$

$$= 0$$

Therefore $\dfrac{1}{\sqrt{54}}(2i + j - 7k)$ is perpendicular to $9i - 4j + 2k$.

7

The Trapezium Rule for six ordinates is:

$$\int_0^{0.5} f(x)\,dx \approx h[\tfrac{1}{2}(y_0 + y_5) + y_1 + y_2 + y_3 + y_4]$$

We are given $h = 0.1$ as the distance between each x-value and $y_0 = 0.1$, $y_1 = 0.23$, $y_2 = 0.45$, $y_3 = 0.52$, $y_4 = 0.44$ and $y_5 = 0.21$.

Therefore

$$\int_0^{0.5} f(x)\,dx \approx 0.1[\tfrac{1}{2}(0.1 + 0.21) + 0.23 + 0.45$$
$$+ 0.52 + 0.44]$$
$$\approx 0.1(0.155 + 0.23 + 0.45 + 0.52 + 0.44)$$
$$\approx 0.1(1.795)$$
$$\approx 0.1795$$

Therefore

$$\int_0^{0.5} f(x)\,dx \approx 0.18 \quad \text{to 2 decimal places}$$

8

Let $\dfrac{1}{r(r+2)} \equiv \dfrac{A}{r} + \dfrac{B}{r+2}$

$\Rightarrow \quad \dfrac{1}{r(r+2)} \equiv \dfrac{A(r+2) + Br}{r(r+2)}$

If we compare the numerators then

$$1 \equiv A(r+2) + Br \qquad [1]$$

To find A we will make B 'disappear' by putting $r = 0$ in [1]

$\Rightarrow \quad 1 = A(0 + 2) + B(0)$

$\Rightarrow \quad 1 = 2A$

$\Rightarrow \quad A = \tfrac{1}{2}$

To find B we will make A 'disappear' by putting $r = -2$ in [1]

$\Rightarrow \quad 1 = A(-2 + 2) + B(-2)$

$\Rightarrow \quad 1 = -2B$

$\Rightarrow \quad B = -\tfrac{1}{2}$

Hence

$$\frac{1}{r(r+2)} \equiv \frac{\tfrac{1}{2}}{r} + \frac{-\tfrac{1}{2}}{r+2} \equiv \frac{1}{2r} - \frac{1}{2(r+2)}$$

$$\Rightarrow \quad \sum_{r=1}^{n} \frac{1}{r(r+2)} \equiv \sum_{r=1}^{n} \left(\frac{1}{2r} - \frac{1}{2(r+2)} \right)$$

To sum this series we substitute $r = 1, 2, 3, \ldots$ and list the fractions as follows:

r	$\dfrac{1}{2r}$	$\dfrac{1}{2(r+2)}$
1	$\frac{1}{2}$	$-\frac{1}{6}$
2	$\frac{1}{4}$	$-\frac{1}{8}$
3	$\frac{1}{6}$	$-\frac{1}{10}$
4	$\frac{1}{8}$	.
5	$\frac{1}{10}$	.
.	.	.
.	.	.
.	.	.
$n-2$	$\dfrac{1}{2(n-2)}$	$-\dfrac{1}{2n}$
$n-1$	$\dfrac{1}{2(n-1)}$	$-\dfrac{1}{2(n+1)}$
n	$\dfrac{1}{2n}$	$-\dfrac{1}{2(n+2)}$

On summation all the fractions cancel apart from the first two in the middle column and the last two in the end column.

Therefore

$$\sum_{r=1}^{n} \frac{1}{r(r+2)} \equiv \sum_{r=1}^{n}\left(\frac{1}{2r} - \frac{1}{2(r+1)}\right)$$

$$= \frac{1}{2} + \frac{1}{4} - \frac{1}{2(n+1)} - \frac{1}{2(n+2)}$$

$$= \frac{3}{4} - \frac{1}{2(n+1)} - \frac{1}{2(n+2)}$$

$$= \frac{3(n+1)(n+2) - 2(n+2) - 2(n+1)}{4(n+1)(n+2)}$$

$$= \frac{3(n^2 + 2n + n + 2) - 2n - 4 - 2n - 2}{4(n+1)(n+2)}$$

$$= \frac{3(n^2 + 3n + 2) - 4n - 6}{4(n+1)(n+2)}$$

$$= \frac{3n^2 + 9n + 6 - 4n - 6}{4(n+1)(n+2)}$$

$$= \frac{3n^2 + 5n}{4(n+1)(n+2)}$$

Therefore

$$\sum_{r=1}^{n} \frac{1}{r(r+2)} = \frac{n(3n+5)}{4(n+1)(n+2)}$$

Let $f(x) \equiv px^3 - 11x^2 + qx + 4$

By the *Factor Theorem* if $f(x)$ is *exactly* divisible by $(x - 1)$ then $f(1) = 0$.

Therefore

$$f(1) = p - 11 + q + 4 = 0$$

$\Rightarrow \quad p + q - 7 = 0$

$\Rightarrow \quad\quad p + q = 7 \hfill [1]$

By the *Remainder Theorem* if $f(x)$ has a *remainder* of 70 when divided by $(x - 3)$ then $f(3) = 70$.

Therefore

$$f(3) = 27p - 99 + 3q + 4 = 70$$

$\Rightarrow \quad 27p + 3q - 95 = 70$

$\Rightarrow \quad\quad 27p + 3q = 165$

$\Rightarrow \quad\quad\quad 9p + q = 55 \hfill [2]$

Subtracting [1] from [2] gives

$$8p = 48$$

$\Rightarrow \quad\quad p = 6$

Substituting $p = 6$ into [1] gives

$$6 + q = 7$$

$\Rightarrow \quad\quad q = 1$

Therefore $f(x) = 6x^3 - 11x^2 + x + 4$ and has $(x - 1)$ as one of its factors.

We now divide f(x) by $(x - 1)$ to obtain the quadratic expression necessary to give the other two factors.

$$
\begin{array}{r}
6x^2 - 5x - 4 \\
x - 1 \overline{\smash{\big)}\, 6x^3 - 11x^2 + x + 4} \\
\underline{6x^3 - 6x^2} \\
-5x^2 + x \\
\underline{-5x^2 + 5x} \\
-4x + 4 \\
\underline{-4x + 4} \\
0
\end{array}
$$

$\Rightarrow \quad 6x^3 - 11x^2 + x + 4 \equiv (x - 1)(6x^2 - 5x - 4)$

Therefore

$$6x^3 - 11x^2 + x + 4 \equiv (x - 1)(2x + 1)(3x - 4)$$

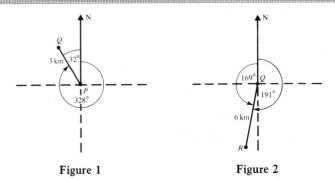

Figure 1   Figure 2

Figures 1 and 2 represent the information given. If we combine these into a single diagram we produce Figure 3.

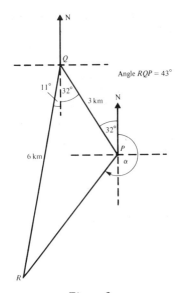

Figure 3

Using the Cosine Rule in triangle RPQ

$$(PR)^2 = 6^2 + 3^2 - 2(3)(6) \cos 43°$$
$$= 36 + 9 - 36 \cos 43°$$
$$= 45 - 36 \cos 43°$$
$$\Rightarrow \quad (PR)^2 = 18.67$$
$$\Rightarrow \quad PR = \sqrt{\{18.67\}} = 4.3 \text{ km} \quad \text{to 1 decimal place}$$

Therefore, the distance from R to P is 4.3 km.

To find the bearing of R from P we work out the value of angle α in Figure 3.

$$\alpha = 360° - (Q\widehat{P}R + 32°) \qquad [1]$$

We must therefore find angle QPR.

To do this we use the Sine Rule

$$\frac{\sin Q\widehat{P}R}{RQ} = \frac{\sin R\widehat{Q}P}{RP}$$

$$\Rightarrow \quad \frac{\sin Q\widehat{P}R}{6} = \frac{\sin 43°}{4.3}$$

$$\Rightarrow \quad \sin Q\widehat{P}R = \frac{6 \times \sin 43°}{4.3} = 0.9516$$

$$\Rightarrow \quad Q\widehat{P}R = 72° \quad \text{or} \quad 108° \quad \text{to the nearest degree}$$

Only *one* of these angles can be correct.

Suppose

$$Q\widehat{P}R = 72°$$

Then

$$Q\widehat{R}P = 180° - (72° + 43°)$$

$$\text{(since the angles of a triangle add up to } 180°)$$

$$= 180° - 115°$$

$$= 65°$$

If $Q\widehat{R}P$ is 65° then we have produced a triangle in which a side of 3 km is opposite an angle of 65° ($Q\widehat{R}P$) and a *larger side* of 4.3 km is opposite a *smaller angle* of 43° ($R\widehat{Q}P$). Such a triangle is impossible.

Therefore

$$Q\widehat{R}P \neq 65°$$

$$\Rightarrow \quad Q\widehat{P}R \neq 72°$$

$$\Rightarrow \quad Q\widehat{P}R = 108°$$

Therefore the bearing of R from P, α, is

$$360° - (108° + 32°) \quad \text{from [1]}$$

$$= 360° - 140°$$

$$= 220°$$

11

To solve this differential equation we separate dy from dx and collect terms in y and dy on the L.H.S. and terms in x and dx on the R.H.S. Then we will be in a position to integrate both sides.

$$xy\frac{\mathrm{d}y}{\mathrm{d}x} = (1 - x^2)$$

$$\Rightarrow \quad xy\,\mathrm{d}y = (1 - x^2)\,\mathrm{d}x$$

$$\Rightarrow \quad y\,\mathrm{d}y = \frac{(1 - x^2)}{x}\mathrm{d}x$$

$$\Rightarrow \quad \int y\,\mathrm{d}y = \int \frac{1 - x^2}{x}\mathrm{d}x \qquad [1]$$

In order to integrate the R.H.S. we rewrite $\dfrac{1-x^2}{x}$ as

$$\frac{1}{x} - \frac{x^2}{x} = \frac{1}{x} - x$$

Therefore equation [1] can be rewritten as

$$\int y\,dy = \int \left(\frac{1}{x} - x\right) dx$$

$$\Rightarrow \quad \int y\,dy = \int \frac{dx}{x} - \int x\,dx$$

$$\Rightarrow \quad \tfrac{1}{2}y^2 = \ln x - \tfrac{1}{2}x^2 + A$$

(where A is a constant of integration)

$$\Rightarrow \quad y^2 = 2\ln x - x^2 + B \quad \text{(where } B = 2A) \qquad [2]$$

We can find the constant of integration by substituting the given values $y = 2$ when $x = 1$ into [2].

Therefore

$$2^2 = 2\ln 1 - 1^2 + B$$

$$\Rightarrow \quad 4 = 2\ln 1 - 1 + B$$

$$\Rightarrow \quad 4 = -1 + B \quad \text{(since } \ln 1 = 0)$$

$$\Rightarrow \quad 5 = B$$

$$\Rightarrow \quad y^2 = 2\ln x - x^2 + 5$$

Therefore

$$y = \sqrt{\{2\ln x - x^2 + 5\}} \quad \begin{array}{l}\text{is the solution to the} \\ \text{differential equation.}\end{array}$$

12

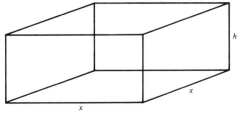

Let Figure 1 represent the box with a square base of side x units and height h units.

Figure 1

The volume, v, of the box is given by

$$V = x \times x \times h$$

$$\Rightarrow \quad V = x^2 h \qquad [1]$$

The surface area, A, of the box is given by

$$A = x^2 + 4xh \qquad [2]$$

where x^2 is the area of the base and xh is the area of each of the four sides.

In order to express A in terms of V and x only we must eliminate h from [2].

From [1]

$$h = \frac{V}{x^2}$$

Substituting this into [2] we obtain the expression:

$$A = x^2 + 4x \left(\frac{V}{x^2} \right)$$

$$\Rightarrow \quad A = x^2 + \frac{4V}{x}$$

The area of the box is a minimum when $\dfrac{dA}{dx} = 0$.

$$A = x^2 + \frac{4V}{x}$$

$$\Rightarrow \quad A = x^2 + 4Vx^{-1}$$

$$\Rightarrow \quad \frac{dA}{dx} = 2x - 4Vx^{-2} \quad (V \text{ is a constant}) \qquad [3]$$

$$\Rightarrow \quad \frac{dA}{dx} = 2x - \frac{4V}{x^2}$$

$$\frac{dA}{dx} = 0 \quad \text{when} \quad 2x - \frac{4V}{x^2} = 0$$

$$\Rightarrow \quad 2x^3 - 4V = 0$$

$$\Rightarrow \quad 2x^3 = 4V$$

$$\Rightarrow \quad x^3 = 2V$$

$$\Rightarrow \quad x = \sqrt[3]{2V} \qquad [4]$$

To confirm this value of x produces a minimum rather than a maximum area we check the sign of $\dfrac{d^2A}{dx^2}$ *when* $x = \sqrt[3]{2V}$

$$\frac{dA}{dx} = 2x - 4Vx^{-2} \quad \text{equation [3]}$$

$$\Rightarrow \quad \frac{d^2A}{dx^2} = 2 + 8Vx^{-3}$$

$$\Rightarrow \quad \frac{d^2A}{dx^2} = 2 + \frac{8V}{x^3}$$

When $x = \sqrt[3]{2V}$

$$\frac{d^2A}{dx^2} = 2 + \frac{8V}{2V}$$

$\Rightarrow \qquad \frac{d^2A}{dx^2} = 6, \quad \text{which is greater than zero}$

Therefore with $x = \sqrt[3]{2V}$, A is a minimum.

To find the ratio $h : x$ we use the relationship

$$h = \frac{V}{x^2} \quad \text{from [1]}$$

$\Rightarrow \qquad h : x \equiv \dfrac{V}{x^2} : x$

$\Rightarrow \qquad \equiv V : x^3$

$\qquad \equiv V : 2V \quad \text{from [4]}$

$\qquad \equiv 1 : 2$

Therefore the ratio of the height of the box to the length of the side of the base is $1 : 2$ when the area is a minimum.

13

We begin by rewriting $(3 + 2x)^k$ *as*

$$3^k \left(1 + \frac{2x}{3}\right)^k$$

To ensure that the Binomial Theorem will give a convergent series in ascending powers of x in the case when k is a fraction

$$\left|\frac{2x}{3}\right| < 1$$

$\Rightarrow \qquad |2x| < 3$

$\Rightarrow \qquad |x| < \frac{3}{2} \quad$ *as stated in the question*

Using the Binomial Theorem

$$(1 + y)^n = 1 + ny + \frac{n(n-1)}{2!}y^2 + n\frac{(n-1)(n-2)}{3!}y^3 + \ldots$$

with $y = \dfrac{2x}{3}$ and $n = k$ we have

$$\left(1 + \frac{2x}{3}\right)^k \equiv 1 + k\left(\frac{2x}{3}\right) + \frac{k(k-1)}{2!}\left(\frac{2x}{3}\right)^2$$
$$+ \frac{k(k-1)(k-2)}{3!}\left(\frac{2x}{3}\right)^3 + \ldots$$

63

$$\Rightarrow \quad (3 + 2x)^k \equiv 3^k \left(1 + \frac{2x}{3}\right)^k$$

$$\equiv 3^k + 3^k k\left(\frac{2x}{3}\right) + 3^k \frac{k(k-1)}{2!}\left(\frac{2x}{3}\right)^2$$

$$+ 3^k \frac{k(k-1)(k-2)}{3!}\left(\frac{2x}{3}\right)^3 + \ldots \quad [1]$$

Therefore the coefficient of x^3 is

$$3^k \frac{k(k-1)(k-2)}{6}\left(\frac{2}{3}\right)^3 \quad \text{from [1]}$$

and the coefficient of x^2 is

$$3^k \frac{k(k-1)}{2}\left(\frac{2}{3}\right)^2 \quad \text{from [1]}$$

Therefore the ratio of these coefficients is

$$\frac{3^k k(k-1)(k-2)}{6}\left(\frac{2}{3}\right)^3 : \frac{3^k k(k-1)}{2}\left(\frac{2}{3}\right)^2$$

$$\equiv \frac{(k-2)}{6}\left(\frac{2}{3}\right)^3 : \frac{1}{2}\left(\frac{2}{3}\right)^2$$

$$\equiv \frac{(k-2)}{6}\left(\frac{2}{3}\right) : \frac{1}{2}$$

$$\equiv \frac{2(k-2)}{18} : \frac{1}{2}$$

$$\equiv \frac{(k-2)}{9} : \frac{1}{2}$$

$$\equiv 2(k-2) : 9$$

When $k = \frac{5}{2}$ this ratio becomes:

$$2(\tfrac{5}{2} - 2) : 9 \equiv 2(\tfrac{1}{2}) : 9$$

$$\equiv 1 : 9$$

Given $k = \frac{5}{2}$ the first term to become negative will be the *first one* to include the bracket $(k - 3)$. This will be

$$3^k \cdot \frac{k(k-1)(k-2)(k-3)}{4!} \cdot \left(\frac{2x}{3}\right)^4$$

which is the term in x^4.

If $\quad x^{r-1} \equiv x^4$

$$r = 5$$

14

$$(z_1 + z_2) = 1 - i + 1 + i\sqrt{3}$$

$$= 2 - i + i\sqrt{3}$$

$$= 2 + (\sqrt{3} - 1)i$$

64

$$\begin{aligned}(z_1 - z_2) &= (1-i)-(1+i\sqrt{3}) \\ &= 1-i-1-i\sqrt{3} \\ &= -i-i\sqrt{3} \\ &= -(1+\sqrt{3})i\end{aligned}$$

We will represent the complex numbers $z_1, z_2, (z_1+z_2)$ and (z_1-z_2) as vectors on the Argand Diagram with $z_1 \equiv \overrightarrow{OP_1}$, $z_2 \equiv \overrightarrow{OP_2}$, $(z_1+z_2) \equiv \overrightarrow{OP_3}$ and $(z_1-z_2) \equiv \overrightarrow{OP_4}$. See Figure 1.

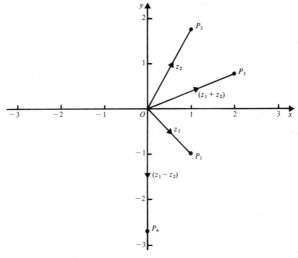

Figure 1

We can also represent a complex number of the form $z = x + iy$ on an Argand Diagram as follows:

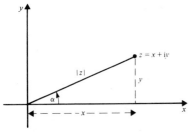

The modulus of z, $|z|$, is represented by the distance from the origin to the point z.

The argument of z, $\arg(z)$, is the angle α, where $-\pi \leqslant \alpha \leqslant \pi$.

(a)

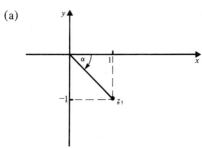

$$|z_1| = \sqrt{1^2 + 1^2} = \sqrt{2} \qquad\qquad [1]$$

$$\arg(z_1) = -\alpha \quad \text{where} \quad \tan\alpha = \frac{1}{1} = 1$$

$$\Rightarrow \quad \alpha = \frac{\pi}{4}$$

$$\Rightarrow \quad \arg(z_1) = -\frac{\pi}{4} \qquad [2]$$

(b)

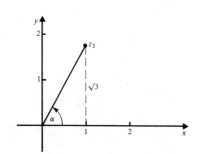

$$|z_2| = \sqrt{1^2 + (\sqrt{3})^2} = \sqrt{1+3} = 2 \qquad [3]$$

$$\arg(z_2) = \alpha = \arctan\frac{\sqrt{3}}{1} = \frac{\pi}{3} \qquad [4]$$

For parts (c) and (d) we use the following four properties of complex numbers.

(i) $|z_1 z_2| = |z_1|\,|z_2|$

(ii) $\arg(z_1 z_2) = \arg(z_1) + \arg(z_2)$

(iii) $\left|\dfrac{z_1}{z_2}\right| = \dfrac{|z_1|}{|z_2|}$

(iv) $\arg\left(\dfrac{z_1}{z_2}\right) = \arg(z_1) - \arg(z_2)$

(c) Using (i)

$$|z_1 z_2| = |z_1|\,|z_2| = \sqrt{2} \times 2 \quad \text{from [1] and [3]}$$
$$\Rightarrow \quad |z_1 z_2| = 2\sqrt{2}$$

Using (ii)

$$\arg(z_1 z_2) = \arg(z_1) + \arg(z_2) = -\frac{\pi}{4} + \frac{\pi}{3}$$

$$\text{from [2] and [4]}$$

$$\Rightarrow \quad \arg(z_1 z_2) = -\frac{3\pi}{12} + \frac{4\pi}{12}$$

$$\Rightarrow \quad \arg(z_1 z_2) = \frac{\pi}{12}$$

(d) Using (iii)

$$\left|\frac{z_1}{z_2}\right| = \frac{|z_1|}{|z_2|} = \frac{\sqrt{2}}{2} \quad \text{from [1] and [3]}$$

$$\Rightarrow \quad \left|\frac{z_1}{z_2}\right| = \frac{1}{2}\sqrt{2}$$

Using (iv)

$$\arg\left(\frac{z_1}{z_2}\right) = \arg(z_1) - \arg(z_2) = -\frac{\pi}{4} - \frac{\pi}{3}$$

$$\text{from [2] and [4]}$$

$$\Rightarrow \quad \arg\left(\frac{z_1}{z_2}\right) = \frac{-3\pi}{12} - \frac{4\pi}{12}$$

$$\Rightarrow \quad \arg\left(\frac{z_1}{z_2}\right) = -\frac{7\pi}{12}$$

15

y is stationary when $\dfrac{dy}{dx} = 0$

$$(1 + x)y = \ln x$$

$$\Rightarrow \quad y = \frac{\ln x}{(1 + x)} \qquad [1]$$

To calculate $\dfrac{dy}{dx}$ we use the Quotient Rule for differentiation

$$\frac{dy}{dx} = \frac{v\dfrac{du}{dx} - u\dfrac{dv}{dx}}{v^2}$$

where $u = \ln x$ and $v = 1 + x$

$$\Rightarrow \quad \frac{du}{dx} = \frac{1}{x} \quad \text{and} \quad \frac{dv}{dx} = 1$$

$$\Rightarrow \quad \frac{dy}{dx} = \frac{(1 + x)\left(\dfrac{1}{x}\right) - \ln x(1)}{(1 + x)^2}$$

$\dfrac{dy}{dx} = 0$ only when the numerator is zero

$$\Rightarrow \quad \frac{dy}{dx} = 0 \quad \text{when} \quad \frac{(1 + x)}{x} - \ln x = 0$$

$$\Rightarrow \quad \ln x = \frac{(1 + x)}{x} \qquad [2]$$

We will sketch the graphs of $y = \ln x$ and $y = \dfrac{(1 + x)}{x}$ on the same diagram.

If they cross at only one point then there will be only one real root of equation [2].

Figure 1 shows the graph of $y = \ln x$.

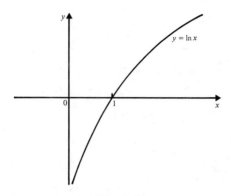

Figure 1

Before we sketch the graph of $y = \dfrac{(1 + x)}{x}$ we rewrite it in the form

$$y = \frac{1}{x} + \frac{x}{x}$$

$$\Rightarrow \quad y = \frac{1}{x} + 1$$

We begin by sketching the graph of $y = \dfrac{1}{x}$ as shown in Figure 2.

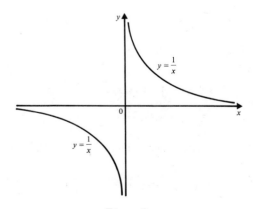

Figure 2

The graph of $y = \dfrac{1}{x} + 1$ is the same as Figure 2 but with each y-value increased by 1 unit. This produces Figure 3, opposite.

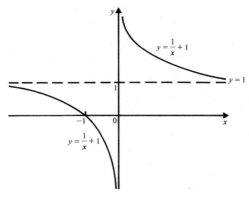

Figure 3

Combining Figures 1 and 3 we produce Figure 4.

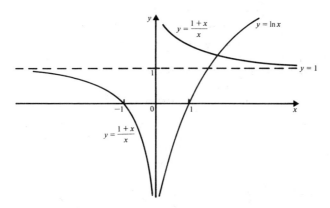

Figure 4

Therefore there is only one solution to $\ln x = \dfrac{(1+x)}{x}$.

We rewrite equation [2] as

$$\ln x - \frac{(1+x)}{x} = 0$$

Let $$f(x) \equiv \ln x - \frac{(1+x)}{x} \qquad\qquad [3]$$

When a curve $y = g(x)$ crosses the x-axis the value of $g(x)$ changes sign from positive to negative, or vice versa.

When $x = 3.5$,

$$f(3.5) = \ln 3.5 - \frac{(1+3.5)}{3.5} \quad \text{from [3]}$$

$$= 1.252\,763 - 1.285\,714\,3$$

$$\Rightarrow \quad f(3.5) = -0.032\,951 \qquad\qquad [4]$$

When $x = 3.8$,

$$f(3.8) = \ln 3.8 - \frac{(1 + 3.8)}{3.8} \quad \text{from [3]}$$

$$= 1.335\,001\,1 - 1.263\,157\,9$$

$$\Rightarrow \quad f(3.8) = 0.071\,843\,2$$

Since the values of $f(3.5)$ and $f(3.8)$ are of different signs there must be a value of x between $x = 3.5$ and $x = 3.8$ for which $f(x) = 0$.

Therefore the only root of $\ln x = \frac{(1 + x)}{x}$ lies in the range $3.5 < x < 3.8$.

The Newton–Raphson procedure states that if $x = 3.5$ is an approximate solution to $f(x) = 0$ then

$$x_1 = 3.5 - \frac{f(3.5)}{f'(3.5)} \qquad [5]$$

is a better approximation.

To apply the procedure we must evaluate $f'(x)$

$$f(x) = \ln x - \frac{(1 + x)}{x} \quad \text{equation [3]}$$

$$= \ln x - \left(\frac{1}{x} + \frac{x}{x} \right)$$

$$= \ln x - \frac{1}{x} - 1$$

$$\Rightarrow \qquad f(x) = \ln x - x^{-1} - 1$$

$$\Rightarrow \qquad f'(x) = \frac{1}{x} + x^{-2}$$

$$\Rightarrow \qquad f'(x) = \frac{1}{x} + \frac{1}{x^2}$$

$$\Rightarrow \qquad f'(3.5) = \frac{1}{3.5} + \frac{1}{(3.5)^2}$$

$$= 0.285\,71 + 0.081\,632$$

$$= 0.367\,34 \qquad [6]$$

Substituting values [4] and [6] into equation [5] gives

$$x_1 = 3.5 - \frac{(-0.032\,951)}{0.367\,34}$$

$$\Rightarrow \qquad x_1 = 3.5 + 0.0897$$

$$\Rightarrow \qquad x_1 = 3.5897$$

$$\Rightarrow \qquad x_1 = 3.59 \quad \text{to 3 significant figures.}$$

An approximation to the corresponding stationary value of y is given by equation [1] with $x = 3.59$.

$$\Rightarrow \quad y = \frac{\ln 3.59}{1 + 3.59}$$

$$\Rightarrow \quad y = \frac{\ln 3.59}{4.59} = 0.278\,46$$

Therefore the stationary value of y is 0.278 to 3 significant figures.

16

The curve is given in a parametric form. We need to investigate x and y for all values of t.

$$x = t^2 - 1 \qquad\qquad\qquad\qquad\qquad [1]$$

$$y = t^3 - t \quad\Rightarrow\quad y = t(t^2 - 1) \qquad\qquad [2]$$

When $t = 0$,

$$x = -1 \quad\text{and}\quad y = 0$$

When $t = 1$,

$$x = 0 \quad\text{and}\quad y = 0$$

Therefore the curve passes through the origin and the point $(-1, 0)$.

For values of $t > 1$ both x and y are positive and the y-values are greater than the x-values (since $y = tx$ from [1] and [2]).

If $0 < t < 1$ then $-1 < x < 0$ (from [1]) and y-values are negative (from [2]).

Therefore for $t \geqslant 0$ the sketch is

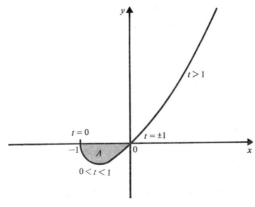

Figure 1

For corresponding negative values of t the x-values are the same as for $t > 0$ and the y-values will be negative.

This information combined with Figure 1 gives Figure 2, overleaf.

71

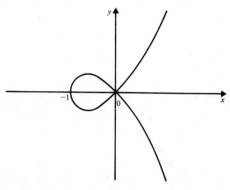

Figure 2

The area enclosed by the loop in Figure 2 is twice the shaded area, A, in Figure 1.

$$A = \int_{-1}^{0} y \, dx, \quad \text{by definition}$$

Before we evaluate this integral we change it into a function of t.

$$y = t^3 - t$$

$$x = t^2 - 1 \quad \Rightarrow \quad \frac{dx}{dt} = 2t$$

$$\Rightarrow \quad dx = 2t \, dt$$

For area A, $0 < t < 1$, therefore the limits are 0 and 1

$$\Rightarrow \quad \int_{-1}^{0} y \, dx \equiv \int_{0}^{1} (t^3 - t)(2t) \, dt$$

$$= \int_{0}^{1} (2t^4 - 2t^2) \, dt$$

$$= \int_{0}^{1} 2t^4 \, dt - \int_{0}^{1} 2t^2 \, dt$$

$$= \left[\tfrac{2}{5} t^5 \right]_{0}^{1} - \left[\tfrac{2}{3} t^3 \right]_{0}^{1}$$

$$= \left(\tfrac{2}{5} - 0 \right) - \left(\tfrac{2}{3} - 0 \right)$$

$$= \tfrac{2}{5} - \tfrac{2}{3}$$

$$\Rightarrow \quad \int_{-1}^{0} y \, dx = -\tfrac{4}{15}$$

(The negative sign corresponds to the area A being under the x-axis.)

Therefore the area enclosed by the loop is

$$2 \times \tfrac{4}{15} = \tfrac{8}{15}$$

72

June 1984

We can solve trigonometric equations involving double angles (i.e. $\sin 2x$ or $\cos 2x$) by use of the double-angle formulae. Normally they reduce to a quadratic equation.

Using $\cos 2x \equiv 1 - 2\sin^2 x$ the original equation becomes

$$1 - 2\sin^2 x = 1 + \sin x$$

$$\Rightarrow \qquad -2\sin^2 x = \sin x$$

$$\Rightarrow \qquad 0 = 2\sin^2 x + \sin x \qquad [1]$$

$$\Rightarrow \qquad \sin x(2\sin x + 1) = 0$$

$$\Rightarrow \qquad \text{either} \quad \sin x = 0 \quad \text{or} \quad 2\sin x + 1 = 0$$

If $\qquad \sin x = 0$

then $\qquad x = 0, \pi, 2\pi$

If $\quad 2\sin x + 1 = 0$

then $\qquad 2\sin x = -1$

$$\Rightarrow \qquad \sin x = -\tfrac{1}{2}$$

$$\Rightarrow \qquad x = \frac{7\pi}{6}, \frac{11\pi}{6}$$

Therefore the solutions for $0 \leqslant x \leqslant 2\pi$ are

$$x = 0, \pi, \frac{7\pi}{6}, \frac{11\pi}{6}, 2\pi$$

N.B. *At [1] we must not divide by $\sin x$ or we shall lose 'half' the solutions.*

If the student must answer at least three of the first five questions we consider the following options.

  (i) She answers 3 of the first 5 questions and 4 of the last 5.
 (ii) She answers 4 of the first 5 questions and 3 of the last 5.
(iii) She answers all of the first 5 questions and 2 of the last 5.

The number of ways she can select option (i) is

$$^5C_3 \times {}^5C_4 = \frac{5!}{2! \times 3!} \times \frac{5!}{1! \times 4!} = 10 \times 5 = 50$$

The number of ways she can select option (ii) is

$$^5C_4 \times {}^5C_3 = \frac{5!}{1! \times 4!} \times \frac{5!}{2! \times 3!} = 5 \times 10 = 50$$

The number of ways she can select option (iii) is

$$1 \times {}^5C_2 = 1 \times \frac{5!}{3! \times 2!} = 1 \times 10 = 10$$

Therefore the total number of ways of selecting 7 questions is

$$50 + 50 + 10 = 110$$

3

If x, y and z are three terms in geometric progression then by
definition $y = xr$ and $z = yr = xr^2$ where r is the common ratio.

If $0.25, a$ and 9 are in geometric progression then

$$a = 0.25r \qquad \qquad [1]$$

and $\quad 9 = ar = 0.25r^2 \quad$ from [1]

$\Rightarrow \qquad 9 = \frac{1}{4}r^2$

$\Rightarrow \qquad 36 = r^2$

$\Rightarrow \qquad r = \pm 6$

$\Rightarrow \qquad a = 0.25(\pm 6) = \pm 1.5 \quad$ from [1]

We are given a as a positive constant

$$\Rightarrow \qquad a = 1.5 \qquad \qquad [2]$$

If u, v and w are three terms in arithmetic progression then by
definition $v = u + d$ and $w = v + d = u + 2d$ where d is the
common difference.

If $0.25, 1.5$ and $9 - b$ are in arithmetic progression then

$$1.5 = 0.25 + d$$

$\Rightarrow \qquad d = 1.5 - 0.25$

$\Rightarrow \qquad d = 1.25$

Therefore the term after 1.5 is

$$1.5 + 1.25 = 2.75$$

$\Rightarrow \qquad 9 - b = 2.75$

$\Rightarrow \qquad -b = -6.25$

$\Rightarrow \qquad b = 6.25$

Therefore $a = 1.5$ and $b = 6.25$.

4

Let $f(x) = x$ and $g(x) = |3x - 8|$.

We shall sketch the two functions f(x) and g(x) on the same axes
and where f(x) lies above g(x) will be the solution to the inequality
$x > |3x - 8|$.

$g(x) = 0$ when $x = \frac{8}{3}$ and by definition $g(x) \geqslant 0$ and is
symmetrical about the line $x = \frac{8}{3}$. We now find the value of
$g(x)$ at two points either side of $x = \frac{8}{3}$ and then sketch the
function $g(x)$.

When $x = 2$

$$g(x) = |6 - 8| = |-2| = 2$$

When $x = 3$

$$g(x) = |9 - 8| = |1| = 1$$

We are now able to sketch the two functions.

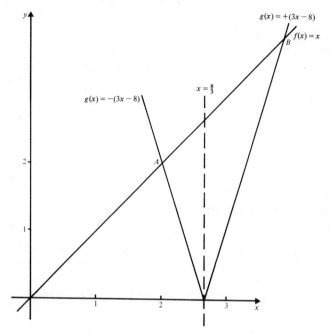

From the diagram $f(x) > g(x)$ for values of x between the two intersection points A and B.

At point A $f(x)$ meets the *negative* branch of $g(x)$

$\Rightarrow$ $x = -(3x - 8)$

$\Rightarrow$ $x = -3x + 8$

$\Rightarrow$ $4x = 8$

$\Rightarrow$ $x = 2$

At point B $f(x)$ meets the *positive* branch of $g(x)$

$\Rightarrow$ $x = 3x - 8$

$\Rightarrow$ $-2x = -8$

$\Rightarrow$ $x = 4$

Therefore $f(x) > g(x)$ and $x > |3x - 8|$ for $\{x : 2 < x < 4\}$.

5

By the *Remainder Theorem* if $P(x)$ has a *remainder* of 2 when divided by $(x - 1)$ then

$$P(1) = 2$$

Therefore

$$P(1) = a + 5 + 2 + b = 2$$

$\Rightarrow$ $a + b + 7 = 2$

$\Rightarrow$ $a + b = -5$ [1]

Similarly, if $P(x)$ has a *remainder* of 5 when divided by $(x - 2)$ then

$$P(2) = 5.$$

Therefore $\quad P(2) = 8a + 20 + 4 + b = 5$

$\Rightarrow \quad 8a + b + 24 = 5$

$\Rightarrow \quad\quad 8a + b = -19$ [2]

Subtracting [2] from [1] gives

$$-7a = 14$$

$\Rightarrow \quad\quad\quad a = -2$

Substituting $a = -2$ into [1] gives

$$-2 + b = -5$$

$\Rightarrow \quad\quad\quad b = -3$

Therefore $a = -2$ and $b = -3$.

6

When we are considering the range of a function it is sometimes useful to visualise a sketch.

When sketching curves of this type we should consider the following:
  (i) does the curve cross either axis (i.e. what happens when $x = 0$ and $y = 0$)?
  (ii) are there any values of x for which the denominator is zero (i.e. are there any vertical asymptotes)?
 (iii) the values of y for large values of x — both positive and negative (i.e. are there any horizontal asymptotes?).
 (iv) how does the curve approach its asymptotes?

It is very important that we explain clearly how the function behaves at each of the steps (i) to (iv) above. We can then mark this information on a diagram (e.g. Figure 1, overleaf) to produce a skeleton sketch. It will then become clear how the sketch may be completed (see Figure 2, overleaf).

(a) Let $y \equiv f(x) \equiv \dfrac{x + 3}{x - 1}$ [1]

    (i) (a) When $x = 0$, $y = -3$

        (b) When $y = 0$, $x + 3 = 0$

            $\Rightarrow$ $x = -3$

    (ii) $x - 1 = 0$ when $x = 1$

        $\Rightarrow$ $x = 1$ is a vertical asymptote

   (iii) (a) As $x \to +\infty$, $y \to 1$ (and is greater than one since $x + 3 > x - 1$).

        (b) As $x \to -\infty$, $y \to 1$ (and is less than one since $|x - 1| > |x + 3|$).

77

(iv) We now check values of x close to either side of the
vertical asymptote, say $x = 1.1$ and $x = 0.9$.

When $x = 1.1$,

$$y = \frac{1.1 + 3}{1.1 - 1} = \frac{4.1}{0.1} = 41 \quad \text{(i.e. large and positive)}$$

When $x = 0.9$,

$$y = \frac{0.9 + 3}{0.9 - 1} = \frac{3.9}{-0.1} = -39 \quad \text{(i.e. large and negative)}$$

This information is used to produce Figure 1.

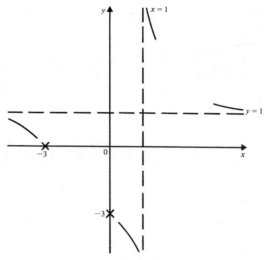

Figure 1

The information in Figure 1 implies that the graph of
$y = \dfrac{x + 3}{x - 1}$ is as shown in Figure 2.

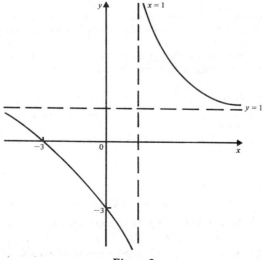

Figure 2

78

From Figure 2 we can see that $f(x)$ takes all values apart from 1.

Therefore the range of $f(x)$ is

$$f(x) \in \mathbb{R}, \qquad f(x) \neq 1$$

(b) $ff(x) = \dfrac{f(x) + 3}{f(x) - 1}$ from [1]

$$= \frac{\dfrac{x + 3}{x - 1} + 3}{\dfrac{x + 3}{x - 1} - 1} \qquad \text{from [1]}$$

$$= \frac{x + 3 + 3(x - 1)}{x + 3 - (x - 1)} \qquad \begin{array}{l}\text{(multiplying numerator and} \\ \text{denominator by } (x - 1))\end{array}$$

$$= \frac{x + 3 + 3x - 3}{x + 3 - x + 1}$$

$$= \frac{4x}{4}$$

Therefore

$$ff(x) = x$$

(c) $ff(x) = x$ implies $f(x)$ is a self-inverse function.

Therefore

$$f^{-1}(x) = f(x)$$

Therefore

$$f^{-1}(x) = \frac{x + 3}{x - 1}, \quad x \in \mathbb{R}, \quad x \neq 1$$

7

Before we integrate to find the area we ensure that the curve $y = x \sin x$ does not cross the x-axis between the limits $x = \dfrac{\pi}{4}$ and $x = \dfrac{\pi}{2}$.

As x and $\sin x$ are both positive for all values of x between $x = \dfrac{\pi}{4}$ and $x = \dfrac{\pi}{2}$, $x \sin x$ will also be positive. Hence for $\dfrac{\pi}{4} \leqslant x \leqslant \dfrac{\pi}{2}$ the curve $y = x \sin x$ does not cross the x-axis.

The required area, A, is given by

$$A = \int_{\pi/4}^{\pi/2} y \, dx$$

$\Rightarrow \qquad A = \int_{\pi/4}^{\pi/2} x \sin x \, dx$

To evaluate this integral we use the formula for Integration by Parts.

$$\int v \frac{du}{dx} dx = uv - \int u \frac{dv}{dx} dx \qquad [1]$$

with

$$v = x \quad \text{and} \quad \frac{du}{dx} = \sin x$$

$$\Rightarrow \quad \frac{dv}{dx} = 1 \quad \text{and} \quad u = -\cos x$$

Substituting into [1] gives

$$A = [(-\cos x)x]_{\pi/4}^{\pi/2} - \int_{\pi/4}^{\pi/2} (-\cos x) \cdot 1 \, dx$$

$$\Rightarrow \quad A = [-x \cos x]_{\pi/4}^{\pi/2} + \int_{\pi/4}^{\pi/2} \cos x \, dx$$

$$= [-x \cos x]_{\pi/4}^{\pi/2} + [\sin x]_{\pi/4}^{\pi/2}$$

$$= \left(-\frac{\pi}{2} \cos \frac{\pi}{2} - -\frac{\pi}{4} \cos \frac{\pi}{4} \right) + \left(\sin \frac{\pi}{2} - \sin \frac{\pi}{4} \right)$$

$$= \left(0 + \frac{\pi}{4} \left(\frac{1}{\sqrt{2}} \right) \right) + \left(1 - \frac{1}{\sqrt{2}} \right)$$

$$= \frac{\pi}{4\sqrt{2}} + 1 - \frac{1}{\sqrt{2}} = \frac{\pi + 4\sqrt{2} - 4}{4\sqrt{2}} = \frac{\pi + 4(\sqrt{2} - 1)}{4\sqrt{2}}$$

Therefore the required area is

$$\frac{\pi + 4(\sqrt{2} - 1)}{4\sqrt{2}} = 0.8483 \quad \text{to 4 significant figures}$$

8

We can represent a complex number of the form $z = x + iy$ on an Argand Diagram as follows.

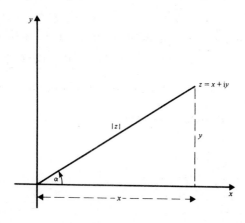

The modulus of z, $|z|$, is represented by the distance from the origin to the point z. The argument of z, $\arg(z)$, is the angle α, where $-\pi \leqslant \alpha \leqslant \pi$.

If $z = x + iy$ then $z^* = x - iy$ is the complex conjugate of z.

(a) $\qquad z_2 = 3 + 4i$

$\Rightarrow \qquad z_2{}^* = 3 - 4i$

$\Rightarrow \qquad z_1 z_2{}^* = (2 + i)(3 - 4i)$

$\qquad\qquad\quad = 6 - 8i + 3i - 4i^2$

$\qquad\qquad\quad = 6 - 5i + 4$

$\Rightarrow \qquad z_1 z_2{}^* = 10 - 5i \qquad\qquad\qquad\qquad\qquad$ [1]

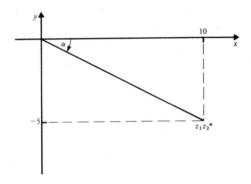

$\qquad |z_1 z_2{}^*| = \sqrt{10^2 + 5^2} = \sqrt{125} = 5\sqrt{5}$

$\qquad \arg(z_1 z_2{}^*) = -\alpha \quad$ where $\quad \tan\alpha = \frac{5}{10} = \frac{1}{2}$

$\Rightarrow \qquad \tan[\arg(z_1 z_2{}^*)] = \tan(-\alpha) = -\tan\alpha = -\frac{1}{2}$

(b) $\quad \dfrac{z_1}{z_2} = \dfrac{2 + i}{3 + 4i}$

To change $\dfrac{z_1}{z_2}$ into the form $x + iy$ we multiply the numerator and the denominator of $\dfrac{z_1}{z_2}$ by the complex conjugate of the denominator. This eliminates i from the denominator.

Therefore

$$\frac{z_1}{z_2} = \frac{z_1 z_2{}^*}{z_2 z_2{}^*} = \frac{10 - 5i}{(3 + 4i)(3 - 4i)} \quad \text{from [1]}$$

$$= \frac{10 - 5i}{9 - 12i + 12i - 16i^2} = \frac{10 - 5i}{9 - 16i^2}$$

$$= \frac{10}{25} - \frac{5}{25}i$$

$\Rightarrow \qquad \dfrac{z_1}{z_2} = \dfrac{2}{5} - \dfrac{1}{5}i$

81

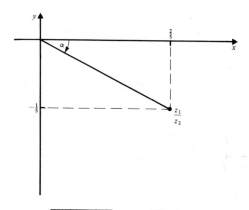

$$\left|\frac{z_1}{z_2}\right| = \sqrt{\left(\frac{2}{5}\right)^2 + \left(\frac{1}{5}\right)^2} = \sqrt{\frac{5}{25}} = \frac{1}{5}\sqrt{5}$$

$$\arg\left(\frac{z_1}{z_2}\right) = -\alpha \quad \text{where} \quad \tan\alpha = \frac{1/5}{2/5} = \frac{1}{2}$$

$$\Rightarrow \quad \tan\left[\arg\left(\frac{z_1}{z_2}\right)\right] = \tan(-\alpha) - \tan\alpha = -\frac{1}{2}$$

9

To show f(x) is always positive we find the range of f(x). We begin by completing the square on $x^2 - 2x + 4$.

$$x^2 - 2x + 4 \equiv (x-1)^2 + 3 \tag{1}$$

$$f(x) \equiv 2 - \frac{3}{x^2 - 2x + 4}$$

$$\Rightarrow \quad f(x) \equiv 2 - \frac{3}{(x-1)^2 + 3} \quad \text{from [1]} \tag{2}$$

Since

$$(x-1)^2 \geqslant 0$$
$$(x-1)^2 + 3 \geqslant 3 \tag{3}$$

Therefore the denominator of the fraction in [2] is greater than or equal to the numerator.

This means that the fraction is always positive and less than or equal to 1,

i.e. $\quad 0 < \dfrac{3}{(x-1)^2 + 3} \leqslant 1$

Therefore

$$1 \leqslant f(x) < 2 \tag{4}$$

Therefore $f(x)$ is always positive.

When sketching curves of this type we should consider the following:

  (i) does the curve cross either axis (i.e. what happens when $x = 0$ and $y = 0$)?

 (ii) are there any values of x for which the denominator is zero (i.e. are there any vertical asymptotes)?

(iii) the values of y for large values of x – both positive and negative (i.e. are there any horizontal asymptotes?).

 (iv) how does the curve approach its asymptotes?

In this type of question the emphasis is more on investigation than algebraic technique. It is very important that we explain clearly how the function behaves at each of the steps (i) to (iv) above. We can then mark this information on a diagram (e.g. Figure 1) to produce a skeleton sketch. It will then become clear how the sketch may be completed (see Figure 2, overleaf).

Let $\quad y = 2 - \dfrac{3}{x^2 - 2x + 4}$

$\Rightarrow \quad y = 2 - \dfrac{3}{(x-1)^2 + 3} \quad$ from [2]

  (i) (a) When $x = 0$, $y = 2 - \dfrac{3}{1+3} = 2 - \tfrac{3}{4} = 1\tfrac{1}{4}$.

       (b) There is no value of x for which $y = 0$ from [4].

 (ii) There are no vertical asymptotes because the denominator is never zero from [3].

(iii) As $x \to \pm\infty$, $y \to 2$ from below from [2] and [4].

(iv) The only asymptote is $y = 2$ and this is approached from below for both large positive and large negative values of x.

This information produces Figure 1.

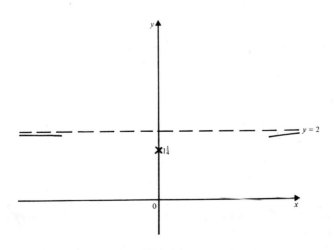

Figure 1

83

The information in Figure 1 together with the fact that the minimum value of y is 1 (from [4]) produces Figure 2.

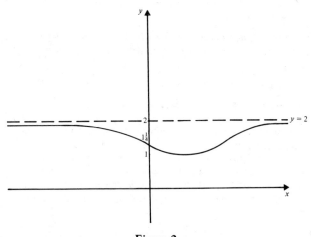

Figure 2

The actual coordinates of the minimum point on the graph $y = f(x)$ are not required. However, it can be seen from [2] that this minimum occurs when $(x-1)^2 = 0$ (i.e. $x = 1$).

Therefore the minimum point is $(1, 1)$.

10

$|y| < 1$ *is stated to ensure that the Binomial Theorem will give a convergent series in ascending powers of y. Apart from this $|y| < 1$ has no relevance within the question.*

Using the Binomial Theorem

$$(1 + x)^n = 1 + nx + \frac{n(n-1)x^2}{2!} + \dots$$

With $x \equiv y$ and $n = \frac{1}{2}$ we have

$$(1 + y)^{1/2} = 1 + \tfrac{1}{2}y + \frac{\frac{1}{2}(-\frac{1}{2})y^2}{2!} + \dots$$

$\Rightarrow \quad (1 + y)^{1/2} = 1 + \tfrac{1}{2}y - \tfrac{1}{8}y^2 + \dots$ [1]

To obtain the expansion of $f(x) \equiv (1 + x^{1/2})^{1/2}$ we substitute $x^{1/2}$ for y in equation [1]

$\Rightarrow \qquad f(x) \equiv (1 + x^{1/2})^{1/2} = 1 + \tfrac{1}{2}(x^{1/2}) - \tfrac{1}{8}(x^{1/2})^2$

$\Rightarrow \qquad f(x) = 1 + \frac{1}{2}x^{1/2} - \frac{1}{8}x + \dots \qquad [2]$

$\Rightarrow \qquad a_0 = 1, \quad a_1 = \frac{1}{2} \quad \text{and} \quad a_2 = -\frac{1}{8}$

$\int_0^{0.01} f(x)\,dx \simeq \int_0^{0.1} (1 + \frac{1}{2}x^{1/2} - \frac{1}{8}x)\,dx \qquad \text{from [2]}$

Since $|y| < 1 \Rightarrow |x^{1/2}| < 1$, *the first 3 terms in the expansion*

of $f(x)$ *will give a good estimate for* $\int_0^{0.01} f(x)\,dx$.)

$\Rightarrow \qquad \int_0^{0.01} f(x)\,dx \simeq [x + \frac{1}{3}x^{3/2} - \frac{1}{16}x^2]_0^{0.01}$

$$\approx \left(0.01 + \frac{(0.01)^{3/2}}{3} - \frac{(0.01)^2}{16} \right) - (0)$$

$$\approx 0.01 + 0.0003 - 0.000\,006\,25$$

$$\approx 0.010\,327$$

Therefore

$$\int_0^{0.01} f(x)\,dx = 0.0103 \qquad \text{to 4 decimal places}$$

11

Let $\quad f(x) \equiv \dfrac{1}{x(x+2)} \equiv \dfrac{A}{x} + \dfrac{B}{(x+2)} \qquad [1]$

Therefore

$$\frac{1}{x(x+2)} \equiv \frac{A(x+2) + Bx}{x(x+2)}$$

If we compare the numerators, then

$$1 \equiv A(x+2) + Bx \qquad [2]$$

To find B we shall make A 'disappear' by putting $x = -2$ in [2]

$\Rightarrow \quad 1 = A(-2 + 2) + B(-2)$

$\Rightarrow \quad 1 = -2B$

$\Rightarrow \quad B = -\frac{1}{2}$

To find A we shall make B 'disappear' by putting $x = 0$ in [2]

$\Rightarrow \quad 1 = A(0 + 2) + B(0)$

$\Rightarrow \quad 1 = 2A$

$\Rightarrow \quad A = \frac{1}{2}$

Hence, substituting $A = \frac{1}{2}$ and $B = -\frac{1}{2}$ into [1] gives

$$f(x) \equiv \frac{1}{x(x+2)} \equiv \frac{\frac{1}{2}}{x} + \frac{-\frac{1}{2}}{(x+2)}$$

$\Rightarrow \qquad f(x) \equiv \dfrac{1}{x(x+2)} \equiv \dfrac{1}{2x} - \dfrac{1}{2(x+2)} \qquad [3]$

(a) $\dfrac{d^4f(x)}{dx^4}$ is the fourth derivative of $f(x)$ with respect to x.

$$f(x) \equiv \frac{1}{2x} - \frac{1}{2(x+2)} \qquad \text{from [3]}$$

$$\Rightarrow \qquad f(x) = \tfrac{1}{2}x^{-1} - \tfrac{1}{2}(x+2)^{-1}$$

$$\Rightarrow \qquad \frac{df(x)}{dx} = -\tfrac{1}{2}x^{-2} + \tfrac{1}{2}(x+2)^{-2}$$

$$\Rightarrow \qquad \frac{d^2f(x)}{dx^2} = x^{-3} - (x+2)^{-3}$$

$$\Rightarrow \qquad \frac{d^3f(x)}{dx^3} = -3x^{-4} + 3(x+2)^{-4}$$

$$\Rightarrow \qquad \frac{d^4f(x)}{dx^4} = 12x^{-5} - 12(x+2)^{-5}$$

Therefore

$$\frac{d^4f(x)}{dx^4} = \frac{12}{x^5} - \frac{12}{(x+2)^5}$$

(b) $$\int_1^3 f(x)\,dx = \int_1^3 \frac{1}{2x}\,dx - \int_1^3 \frac{1}{2(x+1)}\,dx \qquad \text{from [3]}$$

$$= \frac{1}{2}\int_1^3 \frac{dx}{x} - \frac{1}{2}\int_1^3 \frac{dx}{(x+1)}$$

$$= \frac{1}{2}\Big[\ln x\Big]_1^3 - \frac{1}{2}\Big[\ln(x+2)\Big]_1^3$$

$$= \frac{1}{2}\Big\{(\ln 3 - \ln 1) - (\ln 5 - \ln 3)\Big\}$$

$$= \frac{1}{2}\Big\{\ln\left(\frac{3}{1}\right) - \ln\left(\frac{5}{3}\right)\Big\}$$

(by one of the laws of logarithms)

$$= \frac{1}{2}\Big\{\ln\left(\frac{3}{1} \times \frac{3}{5}\right)\Big\}$$

(by one of the laws of logarithms)

$$\Rightarrow \qquad \int_1^3 f(x)\,dx = \frac{1}{2}\ln\frac{9}{5}$$

$$= 0.2939$$

Therefore

$$\int_1^3 f(x)\,dx = 0.2939 \qquad \text{to 4 significant figures.}$$

12

To show that these two lines intersect we must show that for unique values of λ and μ, the lines have a common point. We can do this by equating the i, j and k components of each line.

Let $\quad r_1 = 17i - 9j + 9k + \lambda(3i + j + 5k)$

$\Rightarrow \quad r_1 = 17i - 9j + 9k + 3\lambda i + \lambda j + 5\lambda k$

$\Rightarrow \quad r_1 = (17 + 3\lambda)i + (\lambda - 9)j + (9 + 5\lambda)k$ $\qquad$ [1]

Let $\quad r_2 = 15i - 8j - k + \mu(4i + 3j)$

$\Rightarrow \quad r_2 = 15i - 8j - k + 4\mu i + 3\mu j$

$\Rightarrow \quad r_2 = (15 + 4\mu)i + (3\mu - 8)j - k$ $\qquad$ [2]

Comparing the k components in [1] and [2] gives

$\qquad 9 + 5\lambda = -1$

$\Rightarrow \qquad 5\lambda = -10$

$\Rightarrow \qquad \lambda = -2$ $\qquad$ [3]

Comparing the j components in [1] and [2] gives

$\qquad \lambda - 9 = 3\mu - 8$

$\Rightarrow \quad -2 - 9 = 3\mu - 8 \quad$ from [3]

$\Rightarrow \qquad -11 = 3\mu - 8$

$\Rightarrow \qquad -3 = 3\mu$

$\Rightarrow \qquad -1 = \mu$

Although we have found values of λ and μ which give lines r_1 and r_2 the same j and k components we must check that they also give the same i component.

When $\lambda = -2$ the i component of r_1 is

$\qquad 17 - 6 = 11$

When $\mu = -1$ the i component of r_2 is

$\qquad 15 - 4 = 11$

Therefore when $\lambda = -2$ and $\mu = -1$ the lines r_1 and r_2 have a common point. Therefore the lines r_1 and r_2 intersect.

To find the position vector of this point of intersection we substitute the value $\lambda = -2$ into the line r_1 (or the value $\mu = -1$ into the line r_2).

Substituting $\lambda = -2$ into [1] gives

$\qquad (17 - 6)i + (-2 - 9)j + (9 - 10)k$

$\qquad = 11i - 11j - k$

Therefore the position vector of the point of intersection is

$\qquad r = 11i - 11j - k$

The angle between two lines is defined as the angle between their directions.

Using the definition that the scalar product of two vectors
a and **b** is

$$\mathbf{a} \cdot \mathbf{b} = |\mathbf{a}||\mathbf{b}| \cos \theta \quad \text{(where } \theta \text{ is the angle between them)}$$
[4]

With $\mathbf{a} \equiv 3\mathbf{i} + \mathbf{j} + 5\mathbf{k}$ (the direction vector of $\mathbf{r}_1$)

and $\mathbf{b} \equiv 4\mathbf{i} + 3\mathbf{j}$ (the direction vector of $\mathbf{r}_2$)

we have:

$$\mathbf{a} \cdot \mathbf{b} = (3\mathbf{i} + \mathbf{j} + 5\mathbf{k}) \cdot (4\mathbf{i} + 3\mathbf{j}) = 12 + 3 + 0 = 15$$
$$|\mathbf{a}| = \sqrt{3^2 + 1^2 + 5^2} = \sqrt{35}$$

and $|\mathbf{b}| = \sqrt{4^2 + 3^2} = \sqrt{25} = 5$

Substituting these values into [4] gives

$$15 = (\sqrt{35})(5) \cos \theta$$

$$\Rightarrow \quad \frac{15}{5\sqrt{35}} = \cos \theta$$

$$\Rightarrow \quad \cos \theta = \frac{3}{\sqrt{35}}$$

Therefore the cosine of the acute angle contained by the lines
is $\dfrac{3}{\sqrt{35}}$.

13

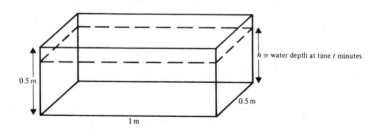

*To find this differential equation we must change the given
information into symbols involving rates of change (e.g.*
$\dfrac{dV}{dt}$ *represents the rate of increase of volume in relation to time).*
*Special care must be taken to ensure that we are consistent in
our use of units.*

At time t minutes the volume of the tank is given by

$$V = 1 \times 0.5 \times h \, \text{m}^3 \quad \text{from Figure 1}$$

$$\Rightarrow \quad V = 0.5h \, \text{m}^3$$
[1]

Also at time t minutes there are 50 litres per minute entering the
tank and $100h$ litres per minute being drained from the tank.

Therefore the overall increase in volume is given by

$$\frac{dV}{dt} = 50 - 100h \quad \ell\,\text{min}^{-1}$$

$$\Rightarrow \quad \frac{dV}{dt} = 0.05 - 0.1h \quad \text{m}^3\,\text{min}^{-1} \quad (\text{since } 1000\,\ell = 1\,\text{m}^3) \quad [2]$$

To find a differential equation involving $\dfrac{dh}{dt}$ we need to form the Chain Rule

$$\frac{dh}{dt} = \frac{dh}{dV} \times \frac{dV}{dt} \tag{3}$$

We can find $\dfrac{dh}{dV}$ by differentiating [1]. This gives

$$\frac{dV}{dh} = 0.5$$

$$\Rightarrow \quad \frac{dh}{dV} = \frac{1}{0.5}$$

$$\Rightarrow \quad \frac{dh}{dV} = 2 \tag{4}$$

Substituting [2] and [4] into [3] gives

$$\frac{dh}{dt} = 2(0.05 - 0.1h)$$

$$\Rightarrow \quad \frac{dh}{dt} = 0.1 - 0.2h$$

$$\Rightarrow \quad 10\frac{dh}{dt} = 1 - 2h$$

To solve the differential equation we shall separate dt from dh by moving dt to the R.H.S. and collect terms in h and dh on the L.H.S. Then we shall integrate each side.

$$10\frac{dh}{dt} = 1 - 2h$$

$$\Rightarrow \quad 10\,dh = (1 - 2h)\,dt$$

$$\Rightarrow \quad \frac{10\,dh}{(1 - 2h)} = dt$$

$$\Rightarrow \quad \int \frac{10\,dh}{(1 - 2h)} = \int dt$$

$$\Rightarrow \quad 10\int \frac{dh}{(1 - 2h)} = \int 1\,dt \tag{5}$$

Since we wish to find the time, T, which elapsed between $h = 0.25$ and $h = 0.375$, we can include this information in [5] to give

$$10\int_{0.25}^{0.375} \frac{dh}{(1 - 2h)} = \int_0^T 1\,dt$$

89

$$\Rightarrow \quad 10[-\tfrac{1}{2}\ln(1-2h)]_{0.25}^{0.375} = [t]_0^T$$

$$\Rightarrow \quad -5[\ln(1-2h)]_{0.25}^{0.375} = [t]_0^T$$

$$\Rightarrow \quad -5[\ln\tfrac{1}{4} - \ln\tfrac{1}{2}] = (T) - (0)$$

$$\Rightarrow \quad -5\ln(\tfrac{1}{4} \times \tfrac{2}{1}) = T \quad \text{by one of the laws of logarithms}$$

$$\Rightarrow \quad -5\ln\tfrac{1}{2} = T$$

$$\Rightarrow \quad -5\ln(2)^{-1} = T$$

$$\Rightarrow \quad 5\ln 2 = T \quad \text{by one of the laws of logarithms}$$

Therefore the tap was opened for $5\ln 2$ minutes.

14

Let $(1 + t^2, 2t - 1)$ be the point on the curve with parameter t, then the gradient of the tangent to the curve at this point is $\dfrac{dy}{dx}$ where

$$\frac{dy}{dx} = \frac{dy}{dt} \times \frac{dt}{dx} \qquad [1]$$

$$x = 1 + t^2$$

$$\Rightarrow \quad \frac{dx}{dt} = 2t$$

$$\Rightarrow \quad \frac{dt}{dx} = \frac{1}{2t} \qquad [2]$$

$$y = 2t - 1$$

$$\Rightarrow \quad \frac{dy}{dt} = 2 \qquad [3]$$

Substituting [2] and [3] into [1] gives

$$\frac{dy}{dx} = 2 \times \frac{1}{2t}$$

$$\Rightarrow \quad \frac{dy}{dx} = \frac{1}{t}$$

The equation of a line of gradient m which passes through the point (a, b) is

$$y - b = m(x - a)$$

Therefore the equation of the tangent passing through $(1 + t^2, 2t - 1)$ with gradient $\dfrac{1}{t}$ is

$$y - (2t - 1) = \frac{1}{t}[x - (1 + t^2)]$$

$$\Rightarrow \quad yt - t(2t-1) = x - (1+t^2)$$

$$\Rightarrow \quad yt - 2t^2 + t = x - 1 - t^2$$

$$\Rightarrow \quad yt + t = x - 1 + t^2$$

$$\Rightarrow \quad yt = x + t^2 - t - 1 \qquad [4]$$

If the point $(2, 1)$ lies on the curve then there is a value of t so that

$$2 = 1 + t^2 \quad and \quad 1 = 2t - 1$$

$$\Rightarrow \quad t^2 = 1 \qquad and \quad 2t = 2$$

$$\Rightarrow \quad t = 1$$

If we substitute $t = 1$ into equation [4] then the equation of the tangent at $A(2, 1)$ is

$$y(1) = x + 1^2 - 1 - 1$$

$$\Rightarrow \quad y = x - 1 \qquad [5]$$

This line passes through $C(6, 5)$ because when $x = 6$ in equation [5], $y = 5$.

If $5y = x + 19$ is a tangent to the curve it must equate with the general form of the tangent given in [4].

Therefore, if

$$5y = x + 19 \equiv yt = x + t^2 - t - 1$$

then $\quad 5 = t$

(since both lines have unit values of x and when $t = 5$,

$$t^2 - t - 1 = 19)$$

Therefore $5y = x + 19$ is a tangent to the curve at the point where $t = 5$.

Therefore the point of contact is

$$(1 + 5^2, 2(5) - 1) = (26, 9)$$

15

Let $\quad y = \sin^{-1}\left(\dfrac{a}{r}\right)$

$$\Rightarrow \quad \sin y = \frac{a}{r} \qquad [1]$$

$$\Rightarrow \quad \sin y = ar^{-1}$$

Differentiating implicitly gives

$$\cos y \, \frac{dy}{dr} = -ar^{-2} = -\frac{a}{r^2}$$

$$\Rightarrow \quad \frac{dy}{dr} = -\frac{a}{r^2 \cos y}$$

$$\Rightarrow \quad \frac{d}{dr}\left[\sin^{-1}\left(\frac{a}{r}\right)\right] = -\frac{a}{r \cos^2 y} \qquad [2]$$

91

To find $\cos y$ in terms of r we use the formula

$$\cos^2 y + \sin^2 y = 1$$

Therefore

$$\cos^2 y = 1 - \sin^2 y$$

$$= 1 - \frac{a^2}{r^2} \quad \text{from [1]}$$

$$\Rightarrow \qquad \cos^2 y = \frac{r^2 - a^2}{r^2}$$

$$\Rightarrow \qquad \cos y = \sqrt{\frac{r^2 - a^2}{r^2}}$$

$$\Rightarrow \qquad \cos y = \frac{\sqrt{r^2 - a^2}}{r}$$

Substituting this value into [2] gives

$$\frac{d}{dr}\left[\sin^{-1}\left(\frac{a}{r}\right)\right] = \frac{-a}{r^2\left(\dfrac{\sqrt{r^2 - a^2}}{r}\right)}$$

Therefore

$$\frac{d}{dr}\left[\sin^{-1}\left(\frac{a}{r}\right)\right] = \frac{-a}{r\sqrt{r^2 - a^2}} \qquad [3]$$

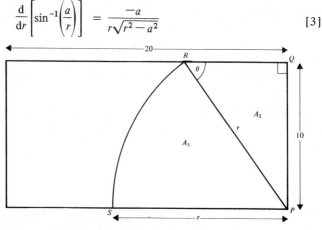

Figure 1

The area, A, which the goat can cover is the area, A_1, of the sector PRS plus the area, A_2, of the triangle PQR.

To find the area of a sector we use the formula

$$\text{area of sector} = \tfrac{1}{2}a^2\theta \qquad [4]$$

(where a is the radius of the sector and θ the angle)

Substituting the values from Figure 1 into formula [4] gives

$$A_1 = \tfrac{1}{2}r^2\theta \qquad [5]$$

From triangle PQR

$$\sin \theta = \frac{10}{r}$$

$$\Rightarrow \qquad \theta = \sin^{-1}\left(\frac{10}{r}\right)$$

Substituting into [5] gives

$$A_1 = \frac{1}{2}r^2 \sin^{-1}\left(\frac{10}{r}\right) \tag{6}$$

$$A_2 = \frac{1}{2}QP \times RQ \quad \text{from Figure 1, opposite}$$

$$= \frac{1}{2}(10)(RQ)$$

$$\Rightarrow \quad A_2 = 5RQ \tag{7}$$

$$RQ^2 = r^2 - 10^2 \quad \text{by Pythagoras' Theorem}$$

$$= r^2 - 100$$

$$\Rightarrow \quad RQ = \sqrt{r^2 - 100}$$

Substituting into [7] gives

$$A_2 = 5\sqrt{r^2 - 100} \tag{8}$$

Adding [8] and [6] gives the required area

$$A = 5\sqrt{r^2 - 100} + \frac{1}{2}r^2 \sin^{-1}\left(\frac{10}{r}\right) \tag{9}$$

To find $\dfrac{dA}{dr}$ we use the fact that

$$\frac{dA}{dr} = \frac{dA_1}{dr} + \frac{dA_2}{dr} \tag{10}$$

To find $\dfrac{dA_1}{dr}$ we use the Product Rule for differentiation

$$\frac{dA_1}{dr} = u\frac{dv}{dr} + v\frac{du}{dr} \tag{11}$$

with

$$u = \frac{1}{2}r^2 \quad \text{and} \quad v = \sin^{-1}\left(\frac{10}{r}\right) \text{from [6]}$$

$$\Rightarrow \quad \frac{du}{dr} = r \quad \text{and} \quad \frac{dv}{dr} = \frac{-10}{r\sqrt{r^2 - 100}}$$

using [3] with $a = 10$

Substituting into [11] gives

$$\frac{dA_1}{dr} = \frac{1}{2}r^2 \times \frac{-10}{r\sqrt{r^2 - 100}} + \sin^{-1}\left(\frac{10}{r}\right) \times r$$

$$\Rightarrow \quad \frac{dA_1}{dr} = \frac{-5r}{\sqrt{r^2\ 100}} + r\sin^{-1}\left(\frac{10}{r}\right) \tag{12}$$

$$\frac{dA_2}{dr} = 5 \times \frac{1}{2}(r^2 - 100)^{-1/2} \times 2r \quad \text{from [8]}$$

$$= 5r(r^2 - 100)^{-1/2}$$

$$\Rightarrow \quad \frac{dA_2}{dr} = \frac{5r}{\sqrt{r^2 - 100}} \tag{13}$$

Substituting results [12] and [13] into [10] gives

$$\frac{\mathrm{d}A}{\mathrm{d}r} = \frac{-5r}{\sqrt{r^2-100}} + r\sin^{-1}\left(\frac{10}{r}\right) + \frac{5r}{\sqrt{r^2-100}}$$

$$\Rightarrow \quad \frac{\mathrm{d}A}{\mathrm{d}r} = r\sin^{-1}\left(\frac{10}{r}\right) \qquad [14]$$

$$A - 100 = 0$$

$$\Rightarrow \quad A = 100$$

This means that the area, A m², to which the goat has access is half the area of the field (which is 20 × 10 = 200 m²).

Therefore

$$5\sqrt{r^2-100} + \tfrac{1}{2}r^2\sin^{-1}\left(\frac{10}{r}\right) = 100 \quad \text{from [9]} \qquad [15]$$

Solving equation [15] finds the value of r which enables the goat to have access to half the area of the field.

Let $\quad f(r) = 5\sqrt{r^2-100} + \tfrac{1}{2}r^2\sin^{-1}\left(\dfrac{10}{r}\right) - 100 \quad$ from [15]

$$[16]$$

The Newton–Raphson procedure states that if $r = 10$ is an approximate solution to $f(r) = 0$ then

$$r_1 = 10 - \frac{f(10)}{f'(10)} \qquad [17]$$

is a better approximation.

$$f(10) = 5\sqrt{10^2-100} + \tfrac{1}{2}(10)^2\sin^{-1}\left(\frac{10}{10}\right) - 100$$

$$\text{from [16]}$$

$$= 0 + 50\sin^{-1}(1) - 100$$

$$= 50\left(\frac{\pi}{2}\right) - 100 \quad \left(\text{since} \quad \sin^{-1}(1) = \frac{\pi}{2}\right)$$

$$\Rightarrow \quad f(10) = 25\pi - 100 \qquad [18]$$

$$f'(r) = \frac{\mathrm{d}\left[5\sqrt{r^2-100} + \tfrac{1}{2}r^2\sin^{-1}\left(\dfrac{10}{r}\right) - 100\right]}{\mathrm{d}r}$$

$$= \frac{\mathrm{d}[A - 100]}{\mathrm{d}r} \quad \text{from [9]}$$

$$= \frac{\mathrm{d}A}{\mathrm{d}r} - \frac{\mathrm{d}[100]}{\mathrm{d}r}$$

$$\Rightarrow \quad f'(r) = r\sin^{-1}\left(\frac{10}{r}\right) \quad \text{from [14]} \left(\text{since} \ \frac{\mathrm{d}[100]}{\mathrm{d}r} = 0\right)$$

$$\Rightarrow \quad f'(10) = 10\sin^{-1}\left(\frac{10}{10}\right)$$

$$= 10\sin^{-1}(1)$$

$$= 10\left(\frac{\pi}{2}\right) \quad \left(\text{since} \quad \sin^{-1}(1) = \frac{\pi}{2}\right)$$

$$\Rightarrow \quad f'(10) = 5\pi \qquad [19]$$

Substituting [18] and [19] into [17] gives

$$r_1 = 10 - \frac{25\pi - 100}{5\pi}$$

$$= \frac{50\pi - (25\pi - 100)}{5\pi}$$

$$= \frac{50\pi - 25\pi + 100}{5\pi}$$

$$= \frac{25\pi + 100}{5\pi}$$

$$= \frac{5\pi + 20}{\pi}$$

$\Rightarrow \quad r_1 = 11.4 \quad$ to 3 significant figures

Therefore when r is approximately equal to 11.4 m the goat has access to half the area of the field.

January 1985

We can solve trigonometric equations involving double angles (i.e. sin 2x or cos 2x) by use of the double-angle formulae. Normally they reduce to a quadratic equation.

Using $\cos 2x \equiv 2\cos^2 x - 1$ the original equation becomes

$$2\cos^2 x - 1 - 3\cos x + 2 = 0$$

$\Rightarrow \qquad 2\cos^2 x - 3\cos x + 1 = 0$

$\Rightarrow \qquad (2\cos x - 1)(\cos x - 1) = 0$

$\Rightarrow \qquad$ either $\quad 2\cos x - 1 = 0 \quad$ or $\quad \cos x - 1 = 0$

$\Rightarrow \qquad$ either $\quad \cos x = \frac{1}{2} \quad$ or $\qquad \cos x = 1$

If $\cos x = \frac{1}{2}$

$$x = \frac{\pi}{3} \text{ or } \frac{5\pi}{3}$$

$\Rightarrow \qquad x = \pm\dfrac{\pi}{3}$

$\Rightarrow \qquad x = \pm\dfrac{\pi}{3}, \quad$ plus multiples of 2π for the general solution

$\Rightarrow \qquad x = 2n\pi \pm \dfrac{\pi}{3} \qquad$ for $\quad n = 0, \pm 1, \pm 2, \ldots$

If $\cos x = 1$

$$x = 0 \text{ or } 2\pi, \qquad \text{for} \quad 0 \leqslant x \leqslant 2\pi$$

$\Rightarrow \qquad x = 0$ or 2π plus multiples of 2π for the general solution

$\Rightarrow \qquad x = 2n\pi, \qquad$ for $\quad n = 0, \pm 1, \pm 2, \ldots$

Therefore the general solution is

$$x = 2n\pi \pm \frac{\pi}{3} \quad \text{and} \quad 2n\pi \qquad \text{for} \quad n = 0, \pm 1, \pm 2, \ldots$$

To find $\displaystyle\sum_{r=1}^{n} \frac{1}{(2r-1)(2r+1)}$ *we split* $\dfrac{1}{(2r-1)(2r+1)}$ *into partial fractions and then add the individual fractions for* $r = 1, \ldots, n$.

Let $\dfrac{1}{(2r-1)(2r+1)} \equiv \dfrac{A}{(2r-1)} + \dfrac{B}{(2r+1)}$ $\qquad$ [1]

Therefore

$$\frac{1}{(2r-1)(2r+1)} \equiv \frac{A(2r+1) + B(2r-1)}{(2r-1)(2r+1)}$$

If we compare the numerators, then

$$1 \equiv A(2r+1) + B(2r-1) \qquad\qquad [2]$$

To find B we shall make A 'disappear' by putting $r = -\frac{1}{2}$ in [2].

$\Rightarrow \quad 1 = A(2(-\frac{1}{2}) + 1) + B(2(-\frac{1}{2}) - 1)$

$\Rightarrow \quad 1 = -2B$

$\Rightarrow \quad B = -\frac{1}{2}$

To find A we shall make B 'disappear' by putting $r = \frac{1}{2}$ in [2].

$\Rightarrow \quad 1 = A(2(\frac{1}{2}) + 1) + B(2(\frac{1}{2}) - 1)$

$\Rightarrow \quad 1 = 2A$

$\Rightarrow \quad A = \frac{1}{2}$

Hence, substituting $A = \frac{1}{2}$ and $B = -\frac{1}{2}$ into [1] gives

$$\frac{1}{(2r - 1)(2r + 1)} \equiv \frac{\frac{1}{2}}{(2r - 1)} + \frac{-\frac{1}{2}}{(2r + 1)}$$

$$\Rightarrow \quad \frac{1}{(2r - 1)(2r + 1)} \equiv \frac{1}{2(2r - 1)} - \frac{1}{2(2r + 1)}$$

$$\Rightarrow \quad \sum_{r=1}^{n} \frac{1}{(2r - 1)(2r + 1)} \equiv \sum_{r=1}^{n} \left(\frac{1}{2(2r - 1)} - \frac{1}{2(2r + 1)} \right)$$

To sum this series we substitute $r = 1, 2, 3, \ldots, n$ and list the fractions as follows

r	$\dfrac{1}{2(2r - 1)}$	$-\dfrac{1}{2(2r + 1)}$
1	$\frac{1}{2}$	$-\frac{1}{6}$
2	$\frac{1}{6}$	$-\frac{1}{10}$
3	$\frac{1}{10}$	$-\frac{1}{14}$
4	$\frac{1}{14}$	.
.	.	.
.	.	.
.	.	.
$n - 1$	$\dfrac{1}{2(2n - 2 - 1)} = \dfrac{1}{2(2n - 3)}$	$-\dfrac{1}{2(2n - 2 + 1)} = -\dfrac{1}{2(2n - 1)}$
n	$\dfrac{1}{2(2n - 1)}$	$-\dfrac{1}{2(2n + 1)}$

On summation all the fractions cancel apart from the top one in the middle column and the bottom one in the end column.

Therefore

$$\sum_{r=1}^{n} \frac{1}{(2r - 1)(2r + 1)} = \frac{1}{2} - \frac{1}{2(2n + 1)}$$

$$= \frac{(2n + 1) - 1}{2(2n + 1)}$$

$$= \frac{2n}{2(2n + 1)}$$

$$= \frac{n}{2n + 1}$$

Therefore

$$\sum_{r=1}^{n} \frac{1}{(2r-1)(2r+1)} = \frac{n}{2n+1}$$

To find $\lim\limits_{n \to \infty} \dfrac{n}{(2n+1)}$ *we rewrite* $\dfrac{n}{2n+1}$ *by dividing both numerator and denominator by n.*

$$\frac{n}{(2n+1)} \equiv \frac{\dfrac{n}{n}}{\dfrac{2n}{n} + \dfrac{1}{n}}$$

$$\Rightarrow \qquad \frac{n}{(2n+1)} \equiv \frac{1}{2 + \dfrac{1}{n}}$$

As $n \to \infty$, $\dfrac{1}{n} \to 0$

$$\Rightarrow \qquad \lim_{n \to \infty} \frac{n}{(2n+1)} \equiv \lim_{n \to \infty} \frac{1}{2 + \dfrac{1}{n}} = \frac{1}{2+0}$$

Therefore

$$\lim_{n \to \infty} \frac{n}{(2n+1)} = \frac{1}{2}$$

3

(a) When considering the roots, α and β, of a quadratic equation we write the equation in the general form of

$$ax^2 + bx + c = 0$$

$$\Rightarrow \qquad x^2 + \frac{b}{a}x + \frac{c}{a} = 0 \qquad\qquad [1]$$

In this case, the sum of the roots

$$\alpha + \beta = -\frac{b}{a} \qquad\qquad [2]$$

and the product of the roots

$$\alpha\beta = \frac{c}{a} \qquad\qquad [3]$$

Rearranging $f(x) = 0$ into the general form, [1], gives

$$a - 2x - x^2 = 0$$

$$\Rightarrow \qquad -x^2 - 2x + a = 0$$

$$\Rightarrow \qquad x^2 + 2x - a = 0$$

100

Let α and $(\alpha + 3)$ be the roots of the equation $f(x) = 0$.
Therefore

$$\alpha + (\alpha + 3) = \frac{-2}{1} = -2 \quad \text{from [2]} \quad (\text{with } \beta \equiv \alpha + 3)$$

$\Rightarrow \qquad 2\alpha + 3 = -2$

$\Rightarrow \qquad 2\alpha = -5$

$\Rightarrow \qquad \alpha = -2\frac{1}{2}$ \hfill [4]

and $\quad \alpha(\alpha + 3) = \frac{-a}{1} = -a \quad \text{from [3]} \quad (\text{with } \beta \equiv \alpha + 3)$

$\Rightarrow \qquad -2\frac{1}{2}(-2\frac{1}{2} + 3) = -a \quad \text{from [4]}$

$\Rightarrow \qquad -2\frac{1}{2}(\frac{1}{2}) = -a$

$\Rightarrow \qquad -1\frac{1}{4} = -a$

$\Rightarrow \qquad a = 1\frac{1}{4}$

Therefore $a = 1\frac{1}{4}$ when the roots differ by 3.

(b) To find the set of values of a for which $f(x) < 0$ we consider the range of $f(x)$.
$f(x)$ is a 'negative' quadratic curve and therefore has a maximum point. For $f(x) < 0$ this maximum point must be below the x-axis.

$$f(x) \equiv a - 2x - x^2 \hfill [5]$$

$\Rightarrow \quad f'(x) = -2 - 2x$

The turning point of $y = f(x)$ occurs when $f'(x) = 0$

i.e. when

$\qquad -2 - 2x = 0$

$\Rightarrow \qquad -2 = 2x$

$\Rightarrow \qquad x = -1$

Therefore, $f(x)$ has a maximum point when $x = -1$.

When $x = -1$,

$$f(-1) = a + 2 - 1 = a + 1$$

Therefore, the maximum point is $(-1, a + 1)$.

Therefore, for $f(x) < 0$

$\qquad a + 1 < 0$

$\Rightarrow \qquad a < -1$

Therefore if $a < -1$, $f(x) < 0$ for all values of x.

4

The curve is given in parametric form. We need to investigate x and y for various values of t.

$$x = t^2 - 2 \quad \text{and} \quad y = 2t$$

When

$$t = 0, \quad x = -2 \quad \text{and} \quad y = 0$$
$$t = 1, \quad x = -1 \quad \text{and} \quad y = 2$$
$$t = 2, \quad x = 2 \quad \text{and} \quad y = 4$$
$$t = -1, \quad x = -1 \quad \text{and} \quad y = -2$$
$$t = -2, \quad x = 2 \quad \text{and} \quad y = -4$$

This information is used to produce Figure 1.

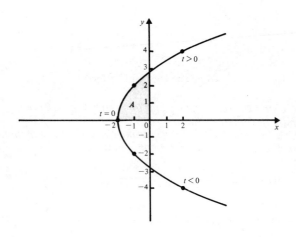

Figure 1

The area of the finite region enclosed by the curve and the y-axis is twice the shaded area, A, in Figure 1.

$$A = \int_{-2}^{0} y \, dx, \quad \text{by definition} \qquad [1]$$

Before we evaluate this integral we change it into a function of t.

$$y = 2t \qquad [2]$$

$$x = t^2 - 2 \Rightarrow \frac{dx}{dt} = 2t \Rightarrow dx = 2t \, dt \qquad [3]$$

To change the limits we evaluate the corresponding t-values for x = −2 and x = 0.

When $x = -2$,

$$-2 = t^2 - 2$$

$\Rightarrow \qquad 0 = t^2$

$\Rightarrow \qquad t = 0 \qquad [4]$

102

When $x = 0$,

$$0 = t^2 - 2$$
$$\Rightarrow \quad 2 = t^2$$
$$\Rightarrow \quad t = \sqrt{2} \qquad\qquad\qquad [5]$$

Substituting [2], [3], [4] and [5] into [1] gives

$$A = \int_0^{\sqrt{2}} 2t(2t)\, dt$$

$$= \int_0^{\sqrt{2}} 4t^2\, dt$$

$$= 4\int_0^{\sqrt{2}} t^2\, dt$$

$$= 4[\tfrac{1}{3}t^3]_0^{\sqrt{2}}$$

$$= 4[\tfrac{1}{3}(\sqrt{2})^3] - 4[0]$$

$$= 4[\tfrac{2}{3}\sqrt{2}]$$

$$\Rightarrow \quad A = \frac{8\sqrt{2}}{3}$$

Therefore the finite region enclosed by the curve and the y-axis is

$$2 \times \frac{8\sqrt{2}}{3} = \frac{16\sqrt{2}}{3}$$

5

Let $f(x) \equiv \dfrac{2}{x-2}$ and $g(x) \equiv \dfrac{1}{x+1}$.

We shall sketch the two functions f(x) and g(x) on the same pair of axes and where f(x) lies below g(x) will be the solution of the inequality

$$\frac{2}{x-2} < \frac{1}{x+1}$$

When sketching curves of this type, we should consider the following:

(i) does the curve cross either axis (i.e. what happens when $x = 0$ and $y - 0$)?

(ii) are there any values of x for which the denominator is zero (i.e. are there any vertical asymptotes)?

(iii) the values of y for large values of x – both positive and negative (i.e. are there any horizontal asymptotes?).

(iv) how does the curve approach its asymptotes?

103

In this type of question the emphasis is more on investigation than algebraic techniques. It is very important that we explain clearly how the function behaves at each of the steps (i) to (iv) above. We can then mark this information on a diagram (e.g. Figures 1 and 2, overleaf) to produce a skeleton sketch. It will then become clear how the sketch may be completed (see Figure 3, p. 106).

For $y = \dfrac{2}{x-2}$

(i) (a) When $x = 0$, $y = -1$.

    (b) There is no value of x for which $y = 0$.

(ii) $x - 2 = 0$ when $x = 2$.

    $\Rightarrow$ $x = 2$ is a vertical asymptote.

(iii) (a) As $x \to +\infty$, $y \to 0$ (and is positive).

    (b) As $x \to -\infty$, $y \to 0$ (and is negative).

    $\Rightarrow$ $y = 0$ is a horizontal asymptote.

(iv) We now check values of x close to each side of the vertical asymptote, say $x = 2.1$ and $x = 1.9$.

When $x = 2.1$,

$$y = \frac{2}{2.1 - 2} = \frac{2}{0.1} = 20 \qquad \text{(i.e. large and positive)}$$

When $x = 1.9$,

$$y = \frac{2}{1.9 - 2} = \frac{2}{-0.1} = -20 \quad \text{(i.e. large and positive)}$$

This information is used to produce Figure 1.

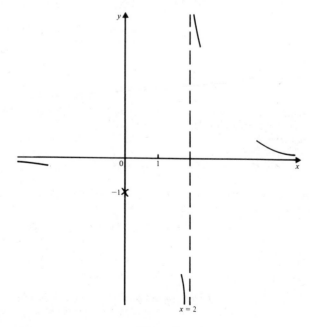

Figure 1

For $y = \dfrac{1}{x+1}$.

(i) (a) When $x = 0$, $y = 1$.

    (b) There is no value of x for which $y = 0$.

(ii) $x + 1 = 0$ when $x = -1$.

    $\Rightarrow$ $x = -1$ is a vertical asymptote.

(iii) (a) As $x \to +\infty$, $y \to 0$ (and is positive).

    (b) As $x \to -\infty$, $y \to 0$ (and is negative).

       $\Rightarrow$ $y = 0$ is a horizontal asymptote.

(iv) We now check values of x close to each side of the vertical asymptote, say $x = -1.1$ and $x = -0.9$.

When $x = -1.1$,

$$y = \frac{1}{-1.1 + 1} = \frac{1}{-0.1} = -10 \quad \text{(i.e. large and negative)}$$

When $x = -0.9$,

$$y = \frac{1}{-0.9 + 1} = \frac{1}{0.1} = 10 \quad \text{(i.e. large and positive)}$$

This information is used to produce Figure 2.

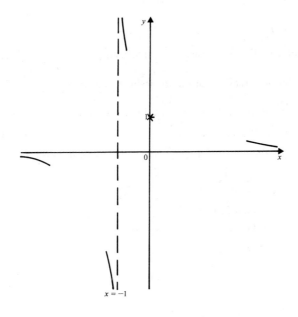

Figure 2

If we combine Figures 1 and 2 we produce Figure 3.

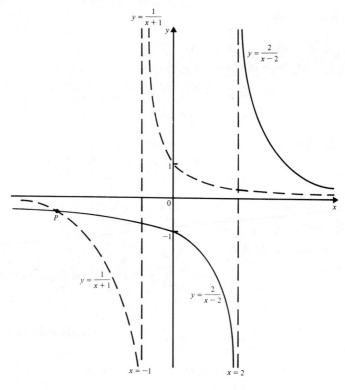

Figure 3

We can find the set of values of x for which $\dfrac{2}{x-2} < \dfrac{1}{x+1}$ by considering the places in Figure 3 where the graph of $y = \dfrac{2}{x-2}$ is below the graph of $y = \dfrac{1}{x+1}$, i.e. for values of x between the two asymptotes $(x = 2$ and $x = -1)$ and to the left of point P. Point P is the point of intersection of the two functions, i.e. when

$$\frac{2}{x-2} = \frac{1}{x+1}$$

Solving this equation gives

$$2(x+1) = x-2$$
$$\Rightarrow \qquad 2x+2 = x-2$$
$$\Rightarrow \qquad x+2 = -2$$
$$\Rightarrow \qquad x = -4$$

Therefore the x-value of the point P is -4.

Hence $\dfrac{2}{x-2} < \dfrac{1}{x+1}$ for

$$\{x: -1 < x < 2 \quad \text{and} \quad x < -4\}$$

6

Each letter of the alphabet can be chosen in 26 ways. Therefore the number of ways of choosing four letters, i.e. three at the beginning and one at the end is

$$26 \times 26 \times 26 \times 26 = (26)^4$$

The number of integers between 1 and 999 inclusive is

999

Therefore the total number of different registration numbers is

$$999 \times (26)^4$$

If all the letters are to be different then

the first letter can be chosen in 26 ways,
the second letter can be chosen in 25 ways,
the third in 24 ways
and the last in 23 ways.

Therefore the total number of ways of choosing the four letters is

$$26 \times 25 \times 24 \times 23 \qquad [1]$$

To find the number of ways of choosing numbers with *different* digits we consider the following options:

(i) a single digit number
(ii) a two digit number
(iii) a three digit number.

In option (i) there are 9 possibilities, i.e. the numbers $1, 2, \ldots, 9$.

In option (ii) there are 9×9 possibilities, i.e. the first digit can be chosen from the numbers $1, 2, \ldots, 9$ and the second digit from the remaining eight digits plus 0.

In option (iii) there are $9 \times 9 \times 8$ possibilities, i.e. the first two digits chosen as in option (ii) and the third digit from the remaining eight digits.

Therefore the total number of ways of choosing numbers with different digits is

$$9 + 9 \times 9 + 9 \times 9 \times 8 = 9 + 81 + 648$$
$$= 738 \qquad [2]$$

Therefore the total number of registrations with letters and digits all different is

$$26 \times 25 \times 24 \times 23 \times 738 \quad \text{from [1] and [2]}$$

7

We are given

$$\frac{x}{(1+x)^2} \equiv \frac{A}{(1+x)^2} + \frac{B}{1+x} \qquad [1]$$

Therefore

$$\frac{x}{(1+x)^2} \equiv \frac{A + B(1+x)}{(1+x)^2}$$

If we compare the numerators, then

$$x \equiv A + B(1+x) \qquad [2]$$

To find A we shall make B 'disappear' by putting $x = -1$ in [2]

$$\Rightarrow \qquad -1 \equiv A + B(1 + -1)$$

$$\Rightarrow \qquad A = -1 \qquad [3]$$

To find B we let $x = 0$ in [2]

$$\Rightarrow \qquad 0 = A + B(1 + 0)$$

$$\Rightarrow \qquad 0 = A + B$$

$$\Rightarrow \qquad 0 = -1 + B \quad \text{from [3]}$$

$$\Rightarrow \qquad B = 1$$

Therefore $A = -1$ and $B = 1$.

Hence, substituting $A = -1$ and $B = 1$ into [1] gives

$$\frac{x}{(1+x)^2} \equiv \frac{-1}{(1+x)^2} + \frac{1}{(1+x)} \qquad [4]$$

From [4]

$$\int_0^1 \frac{x}{(1+x)^2}\,dx \equiv \int_0^1 \left(\frac{-1}{(1+x)^2} + \frac{1}{(1+x)} \right) dx$$

$$= -\int_0^1 \frac{dx}{(1+x)^2} + \int_0^1 \frac{dx}{(1+x)}$$

$$= -\int_0^1 (1+x)^{-2}\,dx + \int_0^1 \frac{dx}{1+x}$$

$$= -[-(1+x)^{-1}]_0^1 + [\ln(1+x)]_0^1$$

$$\Rightarrow \qquad \int_0^1 \frac{x}{(1+x)^2}\,dx = \left[\frac{1}{(1+x)}\right]_0^1 + [\ln(1+x)]_0^1$$

$$= (\tfrac{1}{2} - 1) + (\ln 2 - \ln 1)$$

$$= -\tfrac{1}{2} + \ln 2 \quad \text{since} \quad \ln 1 = 0$$

Therefore

$$\int_0^1 \frac{x}{(1+x)^2}\,dx = \ln 2 - \tfrac{1}{2}$$

8

The gradient of a curve at any point (x, y) is given by

$$\frac{dy}{dx}$$

If this gradient is *directly proportional* to the *product* of x and y then by definition

$$\frac{dy}{dx} = kxy \quad \text{where } k \text{ is a constant}$$

Therefore the differential equation in x and y is

$$\frac{dy}{dx} = kxy \qquad\qquad [1]$$

To find the constant k we substitute the given information that $\frac{dy}{dx} = 4$ at the point $(1, 1)$ into [1] to give

$$4 = k(1)(1)$$

$$\Rightarrow \qquad k = 4$$

Substituting into [1] gives

$$\frac{dy}{dx} = 4xy$$

To solve this differential equation we shall separate dx from dy by moving dx to the R.H.S. and collect terms in y and dy on the L.H.S. Then we shall integrate both sides.

From [1]

$$dy = 4xy\, dx$$

$$\Rightarrow \qquad \frac{dy}{y} = 4x\, dx$$

$$\Rightarrow \qquad \int \frac{dy}{y} = \int 4x\, dx$$

$$\Rightarrow \qquad \int \frac{dy}{y} = 4 \int x\, dx$$

$$\Rightarrow \qquad \ln y = \frac{4x^2}{2} + C \quad (\text{where } C \text{ is a constant of integration})$$

$$\Rightarrow \qquad \ln y = 2x^2 + C \qquad\qquad [2]$$

To find the constant C we substitute the given information that the curve passes through the point $(1, 1)$ (i.e. $x = 1$ when $y = 1$) into [2] to give

$$\ln 1 = 2(1)^2 + C$$

$$\Rightarrow \qquad 0 = 2 + C \quad \text{since} \quad \ln 1 = 0$$

$$\Rightarrow \qquad C = -2 \qquad\qquad [3]$$

Substituting into [2] gives

$$\ln y = 2x^2 - 2$$

$$\Rightarrow \qquad \ln y = 2(x^2 - 1)$$

Therefore $y = e^{2(x^2 - 1)}$ is the solution to the differential equation.

$$y = a^x$$

$\Rightarrow \quad \ln y = \ln a^x \quad \text{(taking logarithms of both sides)}$

$\Rightarrow \quad \ln y = x \ln a \quad \text{(by one of the logarithms)} \qquad [1]$

If x and y are connected by the relationship $y = a^x$ *then plotting* $\ln y$ *against x will give a linear graph (since expression [1] is of the form*

$Y = mx + c, \quad \text{where}$

$Y \equiv \ln y, \quad m \equiv \ln a \quad and \quad c = 0)$

x	2.1	2.8	4.7	6.2	7.3
y	13	32	316	2000	7080
$\ln y$	2.56	3.47	5.76	7.6	8.87

(ln y calculated to 3 significant figures)

Therefore plotting $\ln y$ against x produces the following.

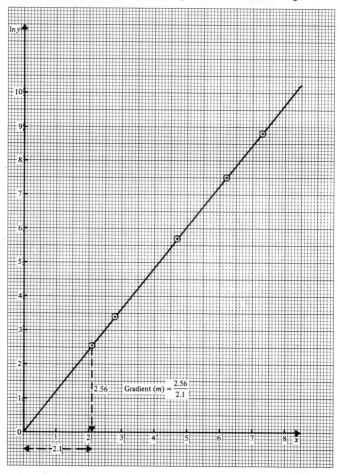

110

To find the value of the constant a we consider the equation of the line.

The equation of a line of gradient m passing through the point (0, c) is

$$Y = mx + c$$

From the graph, $m \approx \dfrac{2.56}{2.1} \approx 1.219$, $Y = \ln y$ and the line passes through (0, 0).

Therefore the equation of the line passing through (0, 0) with gradient 1.219 is

$$\ln y = 1.219x + 0$$

$$\Rightarrow \quad \ln y = 1.219x$$

Comparing this with equation [1] gives

$$\ln a = 1.219$$

$$\Rightarrow \quad a = e^{1.219}$$

$$\Rightarrow \quad a = 3.3839$$

Therefore the constant a is 3.4 to 2 significant figures.

10

Let $f(x) \equiv \sin x - \ln x$ [1]

When a curve, $y = f(x)$, crosses the x-axis the value of $f(x)$ changes sign from positive to negative, or vice versa.

When $x = 2$,

$$f(2) \equiv \sin 2 - \ln 2$$

$$= 0.9093 - 0.6931$$

$$= 0.2162 \qquad [2]$$

When $x = 3$,

$$f(3) \equiv \sin 3 - \ln 3$$

$$= 0.1411 - 1.0986$$

$$= -0.9575$$

Since the values of $f(2)$ and $f(3)$ are of different signs there must be a value of x between $x = 2$ and $x = 3$ for which $f(x) = 0$. Therefore $f(x)$ has a root between $x = 2$ and $x = 3$.

We are given, further, that the root lies between $\dfrac{a}{10}$ and $\dfrac{(a+1)}{10}$

i.e. between $\dfrac{a}{10}$ and $\dfrac{a}{10} + \dfrac{1}{10}$ [3]

As a is an integer, a is in the range

$$20 \leqslant a \leqslant 29 \quad \text{(since the root is between } x = 2 \text{ and } x = 3)$$

$\Rightarrow$ x is in the range

$$2.0 \leqslant x \leqslant 3.0 \quad \text{from [3]}$$

Therefore to find a we evaluate f(x) at intervals of 0.1 (starting with f(2.1)) to see when f(x) changes sign.

$$f(2) = 0.2162 \quad \text{from [2]}$$

$$f(2.1) \equiv \sin 2.1 - \ln 2.1 \quad \text{from [1]}$$

$$= 0.8632 - 0.7419 = 0.1213$$

$$\Rightarrow \quad f(2.1) > 0$$

$$f(2.2) \equiv \sin 2.2 - \ln 2.2 \quad \text{from [1]}$$

$$= 0.8085 - 0.7885 = 0.02 \qquad [4]$$

$$\Rightarrow \quad f(2.2) > 0$$

$$f(2.3) \equiv \sin 2.3 - \ln 2.3 \quad \text{from [1]}$$

$$= 0.7457 - 0.8329 = -0.0872$$

$$\Rightarrow \quad f(2.3) < 0$$

Therefore $f(x)$ has a root between

$$x = 2.2 \quad \text{and} \quad x = 2.3$$

$$\Rightarrow \quad \frac{a}{10} = \frac{22}{10} \quad \text{from [3]}$$

Therefore the integer a is 22.

To find a better estimate for the root we apply the Newton–Raphson procedure *which states that if $x = 2.2$ is an approximate solution to $f(x) = 0$ then*

$$x_1 = 2.2 - \frac{f(2.2)}{f'(2.2)} \qquad [5]$$

is a better approximation.

To apply the procedure we must evaluate $f'(x)$.

$$f(x) = \sin x - \ln x \quad \text{from [1]}$$

$$\Rightarrow \quad f'(x) = \cos x - \frac{1}{x} \qquad [6]$$

$$\Rightarrow \quad f'(2.2) = \cos 2.2 - \frac{1}{2.2}$$

$$= -0.5885 - 0.4545$$

$$\Rightarrow \quad f'(2.2) = -1.043 \qquad [7]$$

Substituting values [4] and [7] into equation [5] gives

$$x_1 = 2.2 - \frac{0.02}{-1.043}$$

$$\Rightarrow \quad x_1 = 2.2 + 0.0192$$

$$\Rightarrow \quad x_1 = 2.2192$$

If $x_1 = 2.2192$ is an approximate solution to $f(x) = 0$ then

$$x_2 = 2.2192 - \frac{f(2.2192)}{f'(2.2192)} \qquad [8]$$

is a better solution.

$$f(2.2192) = \sin(2.2192) - \ln(2.2192) \quad \text{from [1]}$$

$$= 0.7970 - 0.7971$$

$\Rightarrow \quad f(2.2192) = -0.0001$ [9]

$$f'(2.2192) = \cos 2.2192 - \frac{1}{2.2192} \quad \text{from [5]}$$

$$= -0.6039 - 0.4506$$

$\Rightarrow \quad f'(2.2192) = -1.0545$ [10]

Substituting values [9] and [10] into equation [8] gives

$$x_2 = 2.2192 - \frac{-0.0001}{-1.0545}$$

$\Rightarrow \quad x_2 = 2.2192 - 0.000\,095$

$\Rightarrow \quad x_2 = 2.2191$

As the two estimates $x_1 = 2.2192$ and $x_2 = 2.2191$, are the same to three significant figures the estimates value of the root is

$$x = 2.22 \quad \text{to 3 significant figures}$$

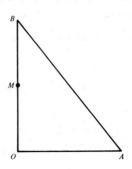

11

Let OAB represent the triangle. If $A\widehat{O}B$ is a right angle then $\overrightarrow{OA}$ will be perpendicular to $\overrightarrow{OB}$ and the scalar product of $\overrightarrow{OA}$ and $\overrightarrow{OB}$ will be zero.

We are given

$$\overrightarrow{OA} = 4i + j - 7k$$ [1]

and $\overrightarrow{OB} = 2i + 6j + 2k$ [2]

$\Rightarrow \quad \overrightarrow{OA}.\overrightarrow{OB} = (4i + j - 7k).(2i + 6j + 2k)$

$$= 8 + 6 - 14 = 0$$

$\Rightarrow \quad \overrightarrow{OA}$ is perpendicular to $\overrightarrow{OB}$

Therefore

$$A\widehat{O}B = 90°$$

The median from A cuts OB at its mid point. The coordinates of B are $(2, 6, 2)$. Therefore the coordinates of M, the mid point of OB, are

$$\left(\frac{0+2}{2}, \frac{0+6}{2}, \frac{0+2}{2}\right) \equiv (1, 3, 1)$$

Therefore the position vector of M is $(i + 3j + k)$.

The vector equation of a line passing through two points with position vectors a and m is given by

$$r = a + \lambda(m - a) \quad \text{(where } \lambda \text{ is a scalar)} \qquad [3]$$

With $\quad a \equiv 4\mathbf{i} + \mathbf{j} - 7\mathbf{k}$ (the position vector of A)

and $\quad m \equiv \mathbf{i} + 3\mathbf{j} + \mathbf{k}$ (the position vector of M)

the vector equation of the median AM is, from [3],

$$\mathbf{r} = (4\mathbf{i} + \mathbf{j} - 7\mathbf{k}) + \lambda[\mathbf{i} + 3\mathbf{j} + \mathbf{k} - (4\mathbf{i} + \mathbf{j} - 7\mathbf{k})]$$

$$\Rightarrow \quad \mathbf{r} = (4\mathbf{i} + \mathbf{j} - 7\mathbf{k}) + \lambda[\mathbf{i} + 3\mathbf{j} + \mathbf{k} - 4\mathbf{i} - \mathbf{j} + 7\mathbf{k}]$$

Therefore a vector equation of the median AM is

$$\mathbf{r} = 4\mathbf{i} + \mathbf{j} - 7\mathbf{k} + \lambda(-3\mathbf{i} + 2\mathbf{j} + 8\mathbf{k})$$

The general vector equation of a plane is

$$\mathbf{r} \cdot \mathbf{n} = p \qquad [4]$$

where r represents a general position vector in the plane; n is a normal vector to the plane and p is a scalar where $\dfrac{p}{|\mathbf{n}|}$ *is the distance of the plane from the origin.*

Since the plane AOB contains the origin, $p = 0$ in [4].

Let $\quad \mathbf{n} = a\mathbf{i} + b\mathbf{j} + c\mathbf{k} \qquad [5]$

be the vector perpendicular to the plane.

Therefore, $\mathbf{n}$ is perpendicular to $\overrightarrow{OA}$

$$\Rightarrow \quad (a\mathbf{i} + b\mathbf{j} + c\mathbf{k}) \cdot (4\mathbf{i} + \mathbf{j} - 7\mathbf{k}) = 0 \quad \text{from [1] and [5]}$$

$$\Rightarrow \quad 4a + b - 7c = 0 \qquad [6]$$

and $\mathbf{n}$ is perpendicular to $\overrightarrow{OB}$

$$\Rightarrow \quad (a\mathbf{i} + b\mathbf{j} + c\mathbf{k}) \cdot (2\mathbf{i} + 6\mathbf{j} + 2\mathbf{k}) = 0 \quad \text{from [2] and [5]}$$

$$\Rightarrow \quad 2a + 6b + 2c = 0 \qquad [7]$$

Multiplying [7] by 2 and subtracting [6] gives

$$4a + 12b + 4c - (4a + b - 7c) = 0$$

$$\Rightarrow \quad 4a + 12b + 4c - 4a - b + 7c = 0$$

$$\Rightarrow \quad 11b + 11c = 0$$

$$\Rightarrow \quad b + c = 0$$

$$\Rightarrow \quad b = -c \qquad [8]$$

Substituting into [7] gives

$$2a + 6(-c) + 2c = 0$$

$$\Rightarrow \quad 2a - 6c + 2c = 0$$

$$\Rightarrow \quad 2a - 4c = 0$$

$$\Rightarrow \quad a - 2c = 0$$

$$\Rightarrow \quad a = 2c \qquad [9]$$

Therefore the vector **n** in [5] may be written as

$$\mathbf{n} = 2c\mathbf{i} - c\mathbf{j} + c\mathbf{k} \quad \text{from [8] and [9]}$$

$\Rightarrow \quad \mathbf{n} = c(2\mathbf{i} - \mathbf{j} + \mathbf{k}) \qquad$ is perpendicular to the plane OAB

$\Rightarrow \quad 2\mathbf{i} - \mathbf{j} + \mathbf{k} \qquad$ is also perpendicular to the plane OAB.

Therefore the vector equation of the plane OAB is

$$\mathbf{r}.(2\mathbf{i} - \mathbf{j} + \mathbf{k}) = 0 \quad \text{from [4], with} \quad p = 0$$

12

Let $\quad z \equiv a + ib$.

We find values of a and b (a and b real) so that

$$z^2 - 5 + 12i = 0$$

i.e. $\quad (a + ib)^2 - 5 + 12i = 0$

$\Rightarrow \qquad (a + ib)^2 = 5 - 12i$

$\Rightarrow \qquad a^2 + 2aib + i^2b^2 = 5 - 12i$

$\Rightarrow \qquad a^2 + 2abi - b^2 = 5 - 12i$

Comparing the real parts gives

$$a^2 - b^2 = 5 \qquad\qquad [1]$$

Comparing the imaginary parts gives

$$2ab = -12$$

$\Rightarrow \qquad ab = -6$

$\Rightarrow \qquad a = \dfrac{-6}{b} \qquad\qquad [2]$

Substituting into [1] gives

$$\left(\frac{-6}{b}\right)^2 - b^2 = 5$$

$\Rightarrow \qquad \dfrac{36}{b^2} - b^2 = 5$

$\Rightarrow \qquad 36 - b^4 = 5b^2$

$\Rightarrow \qquad 36 - b^4 - 5b^2 = 0$

$\Rightarrow \qquad b^4 + 5b^2 - 36 = 0$

$\Rightarrow \qquad (b^2 + 9)(b^2 - 4) = 0$

$\Rightarrow \qquad$ either $b^2 + 9 = 0 \qquad$ or $\quad b^2 - 4 = 0$

$\Rightarrow \qquad\qquad b^2 = -9 \quad$ or $\qquad b^2 = 4$

As b is real, $b^2 = -9$ does not give a possible solution for b.

Therefore,

$$b^2 = 4$$

$\Rightarrow \qquad b = \pm 2$

When $b = 2,$

$\qquad a = -3$ from [2]

When $b = -2,$

$\qquad a = 3$ from [2]

Therefore the values of z so that $z^2 - 5 + 12i = 0$ are

$\qquad -3 + 2i = z_1,$ say

and $\qquad 3 - 2i = z_2,$ say

$\qquad\qquad z_1z_2 = (-3 + 2i)(3 - 2i)$

$\Rightarrow \qquad z_1z_2 = -9 + 6i + 6i - 4i^2$

$\qquad\qquad\quad = -9 + 12i + 4$

Therefore

$\qquad\qquad z_1z_2 = -5 + 12i \qquad\qquad\qquad\qquad [3]$

We can represent a complex number of the form $z = a + ib$ in an Argand Diagram as follows:

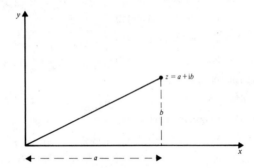

We are asked to draw the Argand Diagram on graph paper. This involves accurately plotting the points $z_1z_2,$ $\dfrac{1}{z_1z_2}$ and $z_1{}^*z_2{}^*$.

(a) $\quad z_1z_2 = -5 + 12i \quad$ equation [3]

(b) *To change $\dfrac{1}{z_1z_2}$ into the form $a + ib$ we multiply the*

numerator and denominator of $\dfrac{1}{z_1z_2}$ by the complex

conjugate of the denominator. This eliminates i from the denominator.

If $z = a + ib$ then $z^* = a - ib$ is the complex conjugate of z.

$$\frac{1}{z_1z_2} = \frac{1}{-5 + 12i} \quad \text{from [3]}$$

$$= \frac{-5 - 12i}{(-5 + 12i)(-5 - 12i)}$$

$$= \frac{-5 - 12i}{25 - 144i^2} = \frac{-5 - 12i}{25 + 144}$$

$$= \frac{-5 - 12i}{169}$$

$$\Rightarrow \quad \frac{1}{z_1 z_2} = -\frac{5}{169} - \frac{12}{169}i \qquad [4]$$

(c) Using $z_1{}^* z_2{}^* \equiv (z_1 z_2)^*$

$$z_1{}^* z_2{}^* = -5 - 12i \quad \text{from [3]} \qquad [5]$$

Therefore, plotting [3], [4] and [5] on graph paper gives

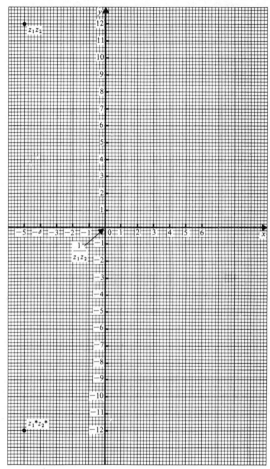

13

(a) $\qquad y(x + y) = 3$

$$\Rightarrow \quad yx + y^2 = 3 \qquad [1]$$

To find $\dfrac{dy}{dx}$ we differentiate [1] implicitly with respect to x.

If $u = yx$ then by the Product Rule for differentiation

$$\frac{du}{dx} = y\frac{d(x)}{dx} + x\frac{d(y)}{dx}$$

$$\Rightarrow \quad \frac{du}{dx} = y + x\frac{dy}{dx} \qquad [2]$$

117

Therefore, using [2], differentiating [1] implicitly with respect to x gives

$$y + x\frac{dy}{dx} + 2y\frac{dy}{dx} = 0$$

$$\Rightarrow \qquad y + \frac{dy}{dx}(x + 2y) = 0$$

$$\Rightarrow \qquad \frac{dy}{dx}(x + 2y) = -y$$

$$\Rightarrow \qquad \frac{dy}{dx} = \frac{-y}{(x + 2y)} \qquad [3]$$

From [1] when $y = 1$

$$x + 1 = 3$$

$$\Rightarrow \qquad x = 2$$

Substituting $y = 1$ and $x = 2$ into [3] gives

$$\frac{dy}{dx} = \frac{-1}{(2 + 2)} = \frac{-1}{4}$$

Therefore the value of $\dfrac{dy}{dx}$ when $y = 1$ is $-\frac{1}{4}$

(b) By the Chain Rule for differentiation

$$\frac{dy}{dx} = \frac{dy}{dt} \times \frac{dt}{dx} \qquad [4]$$

$$x = \frac{1}{(4 - t)^2}$$

$$\Rightarrow \qquad x = (4 - t)^{-2}$$

$$\Rightarrow \qquad \frac{dx}{dt} = -2(4 - t)^{-3}(-1)$$

$$\Rightarrow \qquad \frac{dx}{dt} = \frac{2}{(4 - t)^3}$$

$$\Rightarrow \qquad \frac{dt}{dx} = \frac{(4 - t)^3}{2} \qquad [5]$$

$$y = \frac{t}{4 - t} \qquad [6]$$

To differentiate y with respect to t we use the Quotient Rule for differentiation:

$$\frac{dy}{dt} = \frac{v\dfrac{du}{dt} - u\dfrac{dv}{dt}}{v^2} \qquad [7]$$

where $u = t$ and $v = 4 - t$

$$\Rightarrow \qquad \frac{du}{dt} = 1 \quad \text{and} \quad \frac{dv}{dt} = -1$$

Substituting into [7] gives

$$\frac{dy}{dt} = \frac{(4-t)(1) - t(-1)}{(4-t)^2}$$

$$\Rightarrow \quad \frac{dy}{dt} = \frac{4-t+t}{(4-t)^2}$$

$$\Rightarrow \quad \frac{dy}{dt} = \frac{4}{(4-t)^2} \qquad [8]$$

Substituting [5] and [8] into [4] gives

$$\frac{dy}{dx} = \frac{4}{(4-t)^2} \times \frac{(4-t)^3}{2}$$

$$\Rightarrow \quad \frac{dy}{dx} = 2(4-t) \qquad [9]$$

From [6] when $y = 1$

$$1 = \frac{t}{4-t}$$

$$\Rightarrow \quad 4-t = t$$

$$\Rightarrow \quad 4 = 2t$$

$$\Rightarrow \quad t = 2$$

Substituting into [9] gives

$$\frac{dy}{dx} = 2(4-2) = 4$$

Therefore the value of $\frac{dy}{dx}$ when $y = 1$ is 4.

14

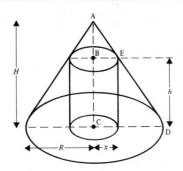

Figure 1

$H \equiv$ height of the solid cone

$R \equiv$ base radius of the cone

$x \equiv$ radius of the cylinder to be made

$h \equiv$ height of the cylinder

The volume, V, of the cylinder is given by

$$V = \pi x^2 h \qquad [1]$$

To find h we consider the following triangle from Figure 1.

119

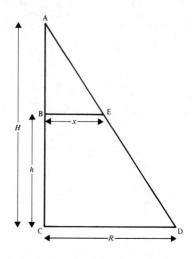

Figure 2

By similar triangles,

$$\frac{CD}{BE} = \frac{AC}{AB}$$

$$\Rightarrow \quad \frac{R}{x} = \frac{H}{H-h} \quad \text{from Figure 2}$$

$$\Rightarrow \quad \frac{R(H-h)}{x} = H$$

$$\Rightarrow \quad RH - Rh = xH$$

$$\Rightarrow \quad -Rh = xH - RH$$

$$\Rightarrow \quad Rh = RH - xH$$

$$\Rightarrow \quad h = \frac{RH - xH}{R}$$

$$\Rightarrow \quad h = \frac{H(R-x)}{R}$$

Substituting into [1] gives

$$V = \frac{\pi x^2 H(R-x)}{R} \qquad [2]$$

Therefore the volume of the cylinder is $\dfrac{\pi H x^2 (R-x)}{R}$.

The volume of the cylinder is a maximum when $\dfrac{dV}{dx} = 0$.

From [2]

$$V = \frac{\pi H x^2 (R-x)}{R}$$

$$\Rightarrow \quad V = \frac{\pi H}{R}(Rx^2 - x^3)$$

$$\Rightarrow \quad \frac{dV}{dx} = \frac{\pi H}{R}(2Rx - 3x^2) \quad \text{(since } \pi, H \text{ and } R \text{ are constants)}$$

$$\frac{dV}{dx} = 0 \quad \text{when} \quad 2Rx - 3x^2 = 0$$

$$\Rightarrow \quad x(2R - 3x) = 0$$

$$\Rightarrow \quad x = 0 \quad \text{or} \quad 2R - 3x = 0$$

$$\Rightarrow \quad x = 0 \quad \text{or} \quad x = \frac{2R}{3}$$

$x = 0$ represents a minimum volume, i.e. zero. Therefore the

Therefore the volume is a maximum when $x = \dfrac{2R}{3}$.

Substituting $x = \dfrac{2R}{3}$ into [2] gives

$$V_{\text{max}} = \frac{\pi\left(\dfrac{2R}{3}\right)^2 H\left(R - \dfrac{2R}{3}\right)}{R}$$

$$= \frac{\pi\left(\dfrac{4R^2}{9}\right)H\left(\dfrac{R}{3}\right)}{R}$$

$$= \frac{4\pi R^2 H}{27}$$

Therefore the maximum possible value of V is $\dfrac{4\pi R^2 H}{27}$.

15

(i) To evaluate $\int (3x + 4)\, e^{2x}\, dx$ we use the formula for

Integration by Parts

$$\int u\frac{dv}{dx}\, dx = uv - \int v\frac{du}{dx}\, dx \qquad [1]$$

with

$$u = 3x + 4 \quad \text{and} \quad \frac{dv}{dx} = e^{2x}$$

$$\Rightarrow \quad \frac{du}{dx} = 3 \qquad \text{and} \quad v = \int e^{2x}\, dx = \tfrac{1}{2}e^{2x}$$

Substituting into [1] gives

$$\int (3x + 4)\, e^{2x}\, dx = (3x + 4)\tfrac{1}{2}e^{2x} - \int \tfrac{1}{2}e^{2x}3\, dx$$

$$= \tfrac{1}{2}(3x + 4)\, e^{2x} - \tfrac{3}{2}\int e^{2x}\, dx$$

$$= \tfrac{1}{2}(3x + 4)\,e^{2x} - \tfrac{3}{4}e^{2x} + C$$

(where C is a constant of integration)

$$= \frac{e^{2x}}{4}\,[2(3x + 4) - 3] + C$$

$$= \frac{e^{2x}}{4}\,(6x + 8 - 3) + C$$

Therefore

$$\int (3x + 4)\,e^{2x}\,dx = \frac{e^{2x}}{4}\,(6x + 5) + C$$

(ii) To evaluate $\displaystyle\int_0^2 \frac{1}{(4 + x^2)^2}\,dx$ *by the substitution* $x = 2\tan\theta$
we change x^2, dx *and the limits into functions of* θ.

$$x = 2\tan\theta \quad \Rightarrow \quad (4 + x^2)^2 \equiv (4 + 4\tan^2\theta)^2$$

$$= 16(1 + \tan^2\theta)^2$$

$$= 16(\sec^2\theta)^2$$

$$\text{since} \quad 1 + \tan^2\theta = \sec^2\theta$$

Therefore

$$(4 + x^2)^2 \equiv 16\sec^4\theta \qquad\qquad [1]$$

$$x = 2\tan\theta \quad \Rightarrow \quad \frac{dx}{d\theta} = 2\sec^2\theta$$

Therefore

$$dx \equiv 2\sec^2\theta\,d\theta \qquad\qquad [2]$$

When $x = 0$,

$$2\tan\theta = 0 \quad \Rightarrow \quad \tan\theta = 0 \quad \Rightarrow \quad \theta = 0 \qquad [3]$$

When $x = 2$,

$$2\tan\theta = 2 \quad \Rightarrow \quad \tan\theta = 1 \quad \Rightarrow \quad \theta = \frac{\pi}{4} \qquad [4]$$

Therefore

$$\int_0^2 \frac{1}{(4 + x^2)^2}\,dx \equiv \int_0^{\pi/4} \frac{2\sec^2\theta\,d\theta}{16\sec^4\theta} \quad \begin{array}{l}\text{from } [1], [2], [3]\\ \text{and } [4]\end{array}$$

$$\equiv \frac{1}{8}\int_0^{\pi/4} \frac{d\theta}{\sec^2\theta}$$

$$\Rightarrow \quad \int_0^2 \frac{1}{(4 + x^2)^2}\,dx \equiv \frac{1}{8}\int_0^{\pi/4} \cos^2\theta\,d\theta \qquad\qquad [5]$$

$$\left(\text{since} \quad \cos^2\theta \equiv \frac{1}{\sec^2\theta}\right)$$

To evaluate $\displaystyle\int \cos^2\theta\,d\theta$ *we make use of the double-angle*
formula

$$\cos 2\theta \equiv 2\cos^2\theta - 1$$

$\Rightarrow \quad \cos 2\theta + 1 \equiv 2 \cos^2\theta$

$\Rightarrow \quad \cos^2\theta \equiv \tfrac{1}{2}(\cos 2\theta + 1)$ [6]

Therefore

$$\int_0^2 \frac{1}{(4 + x^2)^2}\, dx \equiv \frac{1}{8} \int_0^{\pi/4} \tfrac{1}{2}(\cos 2\theta + 1)\, d\theta$$

from [5] and [6]

$$= \frac{1}{16} \int_0^{\pi/4} (\cos 2\theta + 1)\, d\theta$$

$$= \frac{1}{16} \int_0^{\pi/4} \cos 2\theta\, d\theta + \frac{1}{16} \int_0^{\pi/4} 1\, d\theta$$

$$= \frac{1}{16} \left[\tfrac{1}{2} \sin 2\theta \right]_0^{\pi/4} + \frac{1}{16} \left[\theta \right]_0^{\pi/4}$$

$$= \frac{1}{16} \left[\left(\tfrac{1}{2} \sin \frac{\pi}{2} \right) - \left(\tfrac{1}{2} \sin 0 \right) \right]$$

$$+ \frac{1}{16} \left[\frac{\pi}{4} - 0 \right]$$

$$= \frac{1}{16} \left(\frac{1}{2} + \frac{\pi}{4} \right)$$

since $\sin \dfrac{\pi}{2} = 1$ and $\sin 0 = 0$

$$= \frac{1}{16} \left(\frac{2 + \pi}{4} \right)$$

Therefore

$$\int_0^2 \frac{1}{(4 + x^2)^2}\, dx = \frac{2 + \pi}{64}$$

INDEX

Vectors